Trigonometry Basics: Solving Triangles and Trigonometric Functions

TRIGONOMETRY BASICS: SOLVING TRIANGLES AND TRIGONOMETRIC FUNCTIONS

Dr. Mamta Singh

ARCLER
PRESS

www.arclerpress.com

Trigonometry Basics: Solving Triangles and Trigonometric Functions

Dr. Mamta Singh

Arcler Press
224 Shoreacres Road
Burlington, ON L7L 2H2
Canada
www.arclerpress.com
Email: orders@arclereducation.com

ISBN: 978-1-77469-945-4 (Hardcover)

ABOUT THE AUTHOR

Dr Mamta Singh has been associated with Bundelkhand University Jhansi for more than 22 years as a competent , dynamic and responsible faculty member in the mathematics.

She is working as an Associate Professor in Department of Mathematics Sciences and Computer Applications Bundelkhand University, Jhansi. She is awarded with the degree of PhD by Barkatullah University, Bhopal in 2001. She has authored two books and 42 research papers has been published in various National and international journals. She has also attended various conferences and workshops held across the country.

TABLE OF CONTENTS

LIST OF FIGURES

LIST OF TABLES

LIST OF ABBREVIATIONS

2D	Two-Dimensional
3D	Three-Dimensional
AC	Alternating Current
AM	Amplitude Modulation
CAD	Computer-Assisted Design
DFT	Discrete Fourier Transform
DSP	Digital Signal Processing
ETA	Estimating the Time of Arrival
GIS	Geographic Information Systems
GPS	Global Positioning System
MRI	Magnetic Resonance Imaging
ODEs	Ordinary Differential Equations
PDEs	Partial Differential Equations
RFT	Rapid Fourier Transformations
SHM	Simple Harmonic Motion
SNR	Signal-to-Noise Ratio
VR	Virtual Reality

PREFACE

Trigonometry basics involve the study of triangles and trigonometric functions, which play a fundamental role in mathematics and various fields such as physics, engineering, and navigation. When it comes to solving triangles, trigonometry provides methods to determine missing angles or side lengths using the relationships between the angles and sides of a triangle. Trigonometric functions like sine, cosine, and tangent define these relationships and are essential tools for solving problems involving angles and distances. These functions are not only crucial for geometry but also for analyzing periodic phenomena in fields like astronomy and oscillatory motion in physics, making them indispensable in our understanding of the natural world.

This book is comprised of eight chapters, the first chapter introduces the basic concepts of trigonometry. In this opening chapter, readers will gain insights into the historical context and practical applications of trigonometry. It lays the groundwork for subsequent chapters by emphasizing trigonometry's relevance across various fields.

The second chapter provides an in-depth exploration of the Tangent trigonometric function. This chapter not only explains its fundamental properties but also illustrates its practical utility through real-world contexts. The third chapter delves into an extensive examination of Sine and Cosine trigonometric functions. Readers will gain a profound understanding of their behavior and practical significance in solving a wide array of real-world problems.

The fourth chapter deals with the identities and formulas in trigonometry, this chapter equips readers with a comprehensive set of tools for addressing complex trigonometric equations, fostering a deep comprehension of these foundational mathematical concepts. The fifth chapter is about triangles. It offers a comprehensive framework for solving triangles using trigonometric principles, highlighting the practical applications of trigonometry in geometry and beyond.

The sixth chapter expands the discussion to encompass three-dimensional scenarios and angles of varying magnitudes, broadening the reader's skill set and application potential. Chapter seven provides a comprehensive understanding of the intricate relationships between angles and trigonometric functions. Finally, the eighth chapter of the book underscores the contemporary relevance of trigonometry by highlighting its seamless integration with modern tools and techniques. This chapter demonstrates how technology enhances the study and application of trigonometry in today's dynamic world.

The book "Trigonometry Basics: Solving Triangles & Trigonometric Functions" has been meticulously written to provide a solid educational foundation in trigonometry.

Whether you are a student embarking on your mathematical journey or a professional seeking to enhance your analytical skills, this textbook serves as an invaluable resource, promoting a deep understanding of trigonometry and its wide-ranging practical applications.

CHAPTER 1

Introduction to Trigonometry

Contents

1.1. INTRODUCTION

Trigonometry is a branch of mathematics that focuses on the study of relationships and properties of triangles, particularly the angles and sides within them. It plays a crucial role in various fields, including geometry, calculus, physics, engineering, and more. Let's explore the core concepts and components of trigonometry (Webber, 1931).

The following are the fundamental concepts in Trigonometry:

- **Angles:** Angles are fundamental entities in trigonometry. They measure the rotation or inclination between two rays that share a common endpoint (vertex). Angles are typically measured in degrees (°) or radians (rad).
- **Right Triangles:** Trigonometry often starts with right triangles, which have one angle measuring 90 degrees (a right angle). The sides of a right triangle are labeled as the hypotenuse, adjacent (base), and opposite (perpendicular). Trigonometric ratios are defined based on these sides.
- **Trigonometric Functions:** These functions relate the angles of a right triangle to the ratios of its sides. The primary trigonometric functions are:
  - **Sine (sin):** Defined as the ratio of the length of the opposite side to the length of the hypotenuse.
  - **Cosine (cos):** Defined as the ratio of the length of the adjacent side to the length of the hypotenuse.
  - **Tangent (tan):** Defined as the ratio of the length of the opposite side to the length of the adjacent side.

Following are the Key Trigonometric Relationships (Moore & LaForest, 2014):

- **Pythagorean Theorem:** In a right triangle, the square of the length of the hypotenuse (the side opposite to the angle) is equal to the sum of the squares of the lengths of the other two sides. This theorem is essential for solving various trigonometric problems.
- **Trigonometric Identities:** Trigonometric identities are equations involving trigonometric functions that are always true. Some common identities include the Pythagorean identities, reciprocal identities, and quotient identities.
- **Trigonometric Ratios:** These ratios are used to relate the angles and sides of a right triangle. The primary trigonometric ratios are:

$$sin(\theta) = opposite / hypotenuse$$

$$cos(\theta) = adjacent / hypotenuse$$

$$tan(\theta) = opposite / adjacent$$

Trigonometry finds extensive applications in various fields, as illustrated below (Bayeh et al., 2009):

- **Geometry:** It helps solve problems involving angles, triangles, and geometric shapes.
- **Physics:** Trigonometric functions describe the oscillatory motion of waves, the motion of celestial bodies, and the behavior of particles in physics.
- **Engineering:** Engineers use trigonometry to design structures, analyze forces, and solve problems related to circuits and mechanics.
- **Navigation:** Trigonometry is crucial for navigation, including determining positions using latitude and longitude.
- **Computer Graphics:** Trigonometry is used extensively in computer graphics to render images, simulate 3D environments, and create animations.
- **Architecture:** Architects use trigonometry to design buildings and calculate measurements for construction.
- **Surveying and Geography:** Surveyors use trigonometry to measure distances and angles when mapping terrain.

In essence, trigonometry provides a powerful set of mathematical tools for understanding and solving problems related to angles, triangles, and periodic phenomena. It serves as a bridge between geometry and calculus, enabling us to analyze complex relationships in both mathematics and the real world.

1.2. ANGLE

An angle is a geometric figure formed by two rays or line segments that share a common endpoint, known as the vertex. Some commonly used terms in angles are (Gel□fand et al., 2001) (Figure 1.1):

- Terminal side: the final position of the ray after rotation
- Initial side: the original ray

- Vertex: point of rotation
- Negative angle: if the direction of rotation is clockwise
- Positive angle: if the direction of rotation is anticlockwise

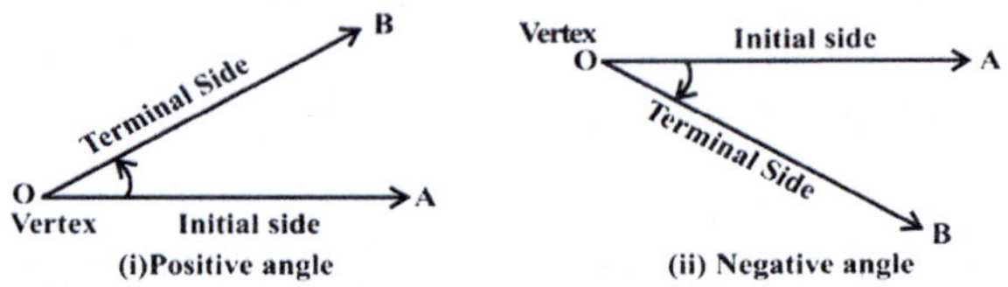

Figure 1.1. Rotation of a ray from an initial point to a terminal point.

Source: Toppr, Creative Commons License.

1.3. QUADRANT

A "quadrant" is a geometric term that refers to one of the four equal divisions or sections into which a plane, space, or object can be divided by two perpendicular lines or axes, typically at right angles to each other. (Birman & Nomizu, 1984) (Figure 1.2).

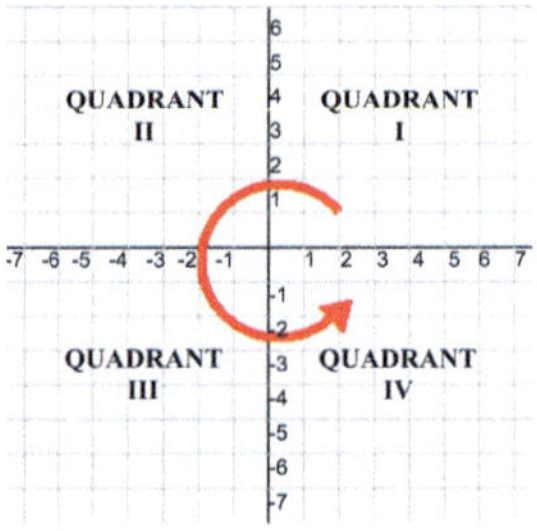

Figure 1.2. Schematic of quadrants.

Source: Varsity Tutors, Creative Commons License.

In Quadrant I, both the x and y coordinates are positive. In Quadrant II, the x coordinate is negative, while the y coordinate is positive. Quadrant III has both coordinates as negative, and in Quadrant IV, the x coordinate is positive while the y coordinate is negative.

Points that fall on an axis, meaning they have at least one coordinate equal to 0, are not considered to be part of any quadrant. Coordinates like (x, 0) are situated on the horizontal x-axis, while coordinates in the form of (0, y) are positioned on the vertical y-axis. (Hirsch et al., 1991).

1.4. MEASUREMENT OF ANGLE

1.4.1. Degree Measure

A full rotation cycle, where the starting and ending points align after either a clockwise or counterclockwise rotation, is divided into 360 units known as degrees. Therefore, if the shift from the starting position to the ending position constitutes 1/360th of a complete rotation, we refer to it as a one-degree angle, symbolized as 1°. (Gerhana et al., 2017) (Figure 1.3).

Following are few examples of angles with their measurements:

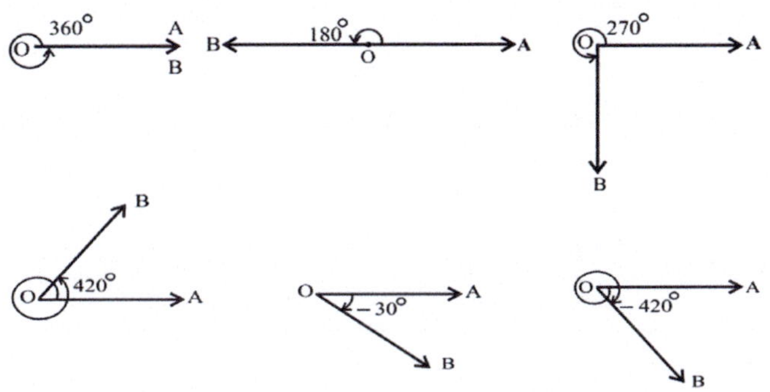

Figure 1.3. Various angles in degree measure.

Source: Toppr, Creative Commons License.

1.4.2. Radian Measure

The angle subtended at the center of a circle by an arc whose length is equal to the radius of the circle is called radian measure (Ungar, 2001) (Figure 1.4):

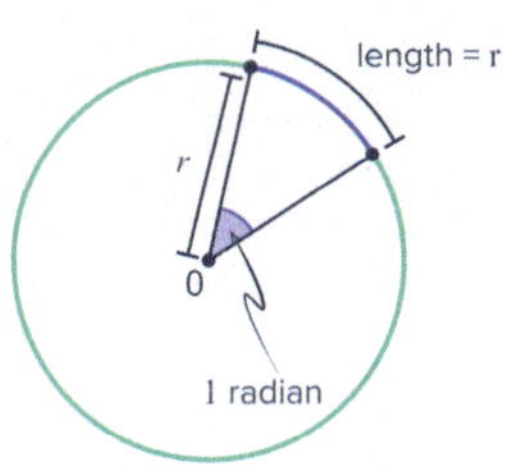

Figure 1.4. Radian measure.

Source: Mathspace, Creative Commons License.

Following are few examples of angles that measure -1 radian, 1.5 radians, and -1.5 radians (Figure 1.5).

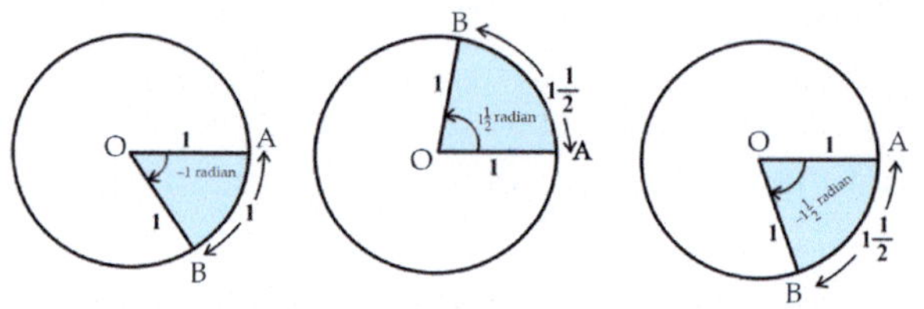

Figure 1.5. Examples of angles that measure in radian.

Source: Toppr, Creative Commons License.

The formula for the circumference of a circle is given by 2πr, where 'r' represents the circle's radius. So, if we have a circle with a radius of 1 unit, its circumference will be 2π units. Hence, one complete revolution of the initial side subtends an angle of 2π radian at the center.

1.4.3. Relation Between Degree and Radian Measure

Consider a circle with a radius 'r.' In accordance with the definition of radians (Figure 1.6):

- An arc with a length equal to 'r' corresponds to an angle of 1 radian.
- An arc with a length of 2πr corresponds to an angle of 2π radians.
- Furthermore, an arc with a length of 2πr corresponds to a full angle of 360 degrees.

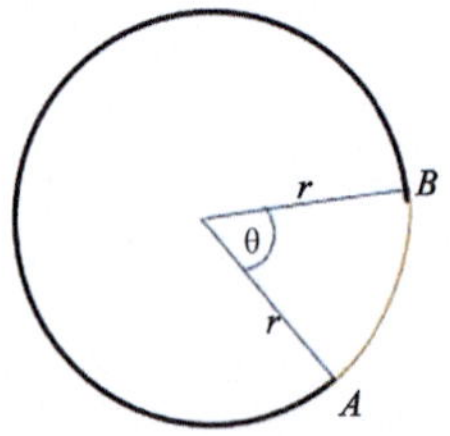

Figure 1.6. Relationship between degree and radian.

Source: Story of Mathematics, Creative Commons License.

So,

π radians = 180°

Or

2 π radians = 360°

1 radian = 180°/ π

1 radian = 180/ 3.1416

Or

1 radian = 57.3°

multiply the number of radians by 180°/ π or 57.3

to convert radians into degrees

Now

360° = 2 π radians

$1^{o} = 2\pi / 360^{o}$

$1^{o} = \pi / 180^{o}$

$1^{o} = 3.1416 / 180^{o}$

1°= 0.01745 radians.

1.5. TRIGONOMETRIC RATIOS

Trigonometric ratios play a crucial role in trigonometry as they establish a relationship between angles and the relative lengths of sides within a right triangle (Blackett & Tall, 1991). The presence of a 90-degree angle in a right triangle gives rise to three fundamental trigonometric ratios: sine (sin), cosine (cos), and tangent (tan).

- The sine function, denoted as sinθ, represents the relationship between the length of the side opposite to the angle θ and the length of the hypotenuse in a right triangle (Bernard et al., 2019):

$$sin\theta = \frac{opposite}{hypotenuse}$$

The trigonometric function cosine of θ (cosθ) can be defined as the ratio of the length of the side adjacent to the angle θ to the length of the hypotenuse:

$$cos\theta = \frac{adjacent}{hypotenuse}$$

- The tangent of an angle θ, denoted as tanθ, is defined as the ratio of the length of the side opposite to the angle to the length of the side adjacent to the angle (Fahrudin & Pramudya, 2019):

$$tan\theta = \frac{opposite}{adjacent}$$

The ratios are fundamental for comprehending the interconnections between angles and sides within a right triangle.

The Pythagorean theorem, which holds a significant position in trigonometry, is a fundamental proposition that establishes a connection between the disciplines of geometry and algebra. It is postulated that within a right triangle, the square of the hypotenuse's length is equivalent to the sum of the squares of the lengths of the remaining two sides (Chambers, 1999). Symbolically:

$$hypotenuse^2 = opposite^2 + adjacent^2$$

The theorem functions as a geometric tool, providing a means to comprehend the interrelationships among the sides of a right triangle and establishing the fundamental basis for more intricate trigonometric principles (Saville et al., 1991).

The unit circle serves as a geometric construct that effectively establishes a connection between angles and trigonometric functions. The unit circle, centered at the origin and has a radius of one, is useful for comprehending the sine and cosine functions for all angles (Moore et al., 2012).

Consider an angle denoted by θ. In the context of the unit circle, the x-coordinate of a point associated with the angle θ denotes the cosine of θ (cos θ), and the y-coordinate indicates the sine of θ (sin θ).

The expression of this link is shown in the equation of the unit circle (Strang, 1986):

$$x^2 + y^2 = 1$$

Trigonometric functions extend their utility beyond right triangles and encompass a wider range of angles by using the unit circle concept (Yang & Sianturi, 2017).

1.6. TRIGONOMETRIC FUNCTIONS

A set of mathematical functions that relate the angles of a right triangle to the ratios of the lengths of its sides are known as trigonometric functions (Figure 1.7).

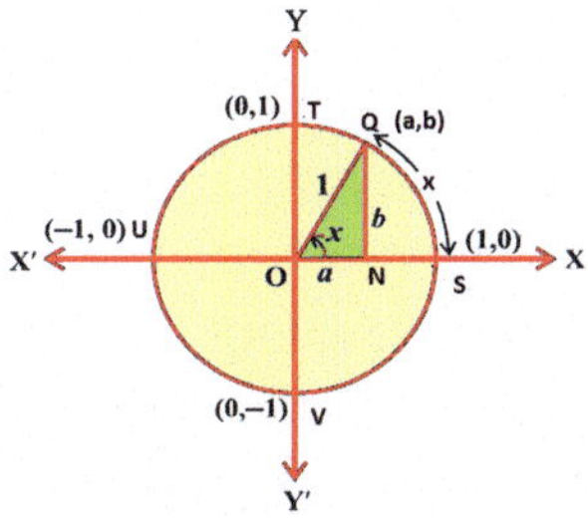

Figure 1.7. Diagram representing a circle with origin 'O.'

Source: Toppr, Creative Commons License.

The diagram depicts a circle with its center labeled as O, situated at the origin of the coordinate axes. Any point on the circle can be denoted as Q(a, b), and the angle SOQ is designated as x radians. The length of the arc from point S to point Q, formed by the angle SOQ, is also equal to x radians. Now, we define cos x=a and sin x = b (Edmunds et al, 2012).

1.6.1. Quadrantal Angles

In the provided figure, it is evident that ONQ forms a right-angled triangle; therefore,

$ON^2 + NQ^2 = OQ^2$

or $a^{2+}b^2 = 1$

or, $\cos^2 x + \sin^2 x = 1$

"This means that every point on the unit circle can be expressed as $\cos^2 x + \sin^2 x$. Now, considering that a complete revolution around the circle corresponds to an angle of 2π radians at the center, we can deduce that ∠ SOT is π/2, ∠ SOU is π, and ∠ SOV is 3π/2.

The angles we've just mentioned are referred to as quadrantal angles because they are exact multiples of π/2. The coordinates for points S, T, U, and V are (1, 0), (0, 1), (-1, 0), and (0, -1) respectively. Now, let's determine the specific values of these quadrantal angles within the circle mentioned earlier:

$$cos\,0° = 1; sin\,0° = 0$$

$$cos\,\pi / 2 = 0; sin\,\pi / 2 = 1$$

$$cos\,\pi = -1; sin\,\pi = 0$$

$$cos\frac{3\pi}{2}=0; sin\frac{3\pi}{2}=-1$$

$$cos\,2\pi=1; sin\,2\pi=0$$

When we take one complete revolution from point Q, we come back to the same starting point Q. Also, the increase or decrease in the value of x by an integral multiple of 2π does not affect the values of sine and cosine functions.

$\sin(2n\pi + x) = \sin x$, $n \in Z$, $\cos(2n\pi + x) = \cos x$, $n \in Z$

1.6.2. Trigonometric Equations

We already know that the values of sin x and cos x repeat after an interval of 2π. Also, the values of tan x repeat after an interval of π. If the equation involves a variable $0 \leq x < 2\pi$, then the solutions are called principal solutions. A general solution is one that involves the integer 'n' and gives all solutions of a trigonometric equation. Also, the character 'Z' is used to denote the set of integers (Figure 1.8).

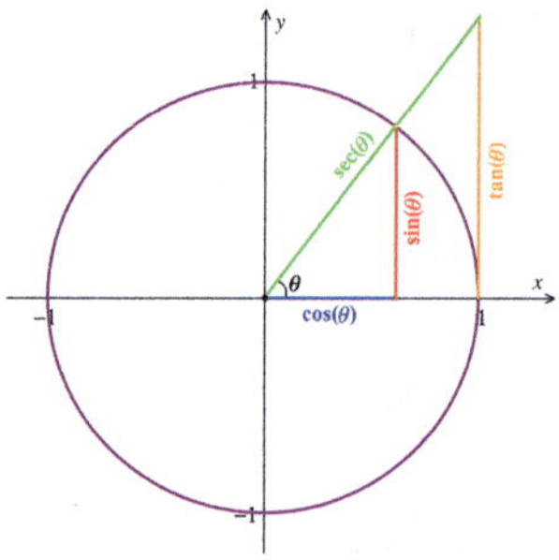

Figure 1.8. Diagram of a triangle and circle for understanding trigonometric equations.

Source: Pixabay, Creative Commons License.

- Examples of Trigonometric Equations:

Let's look at these examples to understand the principal solutions:

Example-1: Find the principal solutions of the equation $sin\,x=\frac{\sqrt{3}}{2}$.

Solution: We know that, $sin\frac{\pi}{3}=\frac{\sqrt{3}}{2}$

Also, $sin\frac{2\pi}{3}=sin\left(\pi-\frac{\pi}{3}\right)$

Now, we know that $sin(\pi-x)=sinx.$

Hence, $sin\frac{2\pi}{3} = sin\frac{\pi}{3} = \frac{\sqrt{3}}{2}$

Therefore, the principal solutions of $sinx =\frac{\sqrt{3}}{2}$ are x = $\frac{\pi}{3}$ and $\frac{2\pi}{3}$.

Example-2: Find the principal solutions of the equation $tanx=-\frac{1}{\sqrt{3}}$

Solution: We know that,

$$\tan\frac{\pi}{6} = \frac{1}{\sqrt{3}}$$

Also, we know that $tan(\pi-x) =- tanx$

Therefore,

$$\tan(\pi-\frac{\pi}{6}) = -\tan\frac{\pi}{6} = -\frac{1}{\sqrt{3}}$$

Further, $tan(2\pi-x) =-tanx$

Therefore,

$$\tan(2\pi-\frac{\pi}{6}) = -\tan\frac{\pi}{6} = -\frac{1}{\sqrt{3}}$$

Hence, the principal solutions of $tanx=-\frac{1}{\sqrt{3}}$ are:

i. $\tan(\pi-\frac{\pi}{6})$ or $= \tan\frac{5\pi}{6}$ and

ii. $\tan(2\pi-\frac{\pi}{6})$ or $\tan\frac{11\pi}{6}$

1.6.3. Solutions for Trigonometric Equations

To start, consider the simple equation sin x = 0. The primary solutions within the interval [0, 2π] are x = 0, π, and 2π. However, it's important to note that if sin x = 0, then x = 0, π, 2π, π, -2π, -3π, and so on, are all valid solutions. Therefore, the general solution for sin x = 0 can be expressed as x = nπ, where $n \in I$ (Table 1.1).

Similarly, general solution for cos x = 0 will be $x=(2n+1)\pi/2, n \in I$, as cos x has a value equal to 0 at $\pi/2, 3\pi/2, 5\pi/2, -7\pi/2, -11\pi/2$ etc.

Table 1.1. Solutions for Trigonometric Equations

Equations	Solutions
$\cos x=0$	$x=(n\pi+\pi/2)$
$\sin x=0$	$x=n\pi$
$\tan x=0$	$x=n\pi$
$\cos x=1$	$x=2n\pi$
$\sin x=1$	$x=(2n\pi+\pi/2)=(4n+1)\pi/2$
$\cos x=\cos\theta$	$x=2n\pi\pm\theta, where\,\theta\in(0,\pi]$
$\sin x=\sin\theta$	$x=n\pi+(-1)n\theta, where\,\theta\in[-\pi/2,\pi/2]$
$\tan x=\tan\theta$	$x=n\pi+\theta, where\,\theta\in(-\pi/2,\pi/2]$
$\cos 2x=\cos 2\theta$	$x=n\pi\pm\theta$

$sin\,2x = sin\,2\theta$	$x = n\pi \pm \theta$
$tan\,2x = tan\,2\theta$	$x = n\pi \pm \theta$

Source: BYJUs, Creative Commons License.

- **Proof of Solutions of Trigonometric Equations:**

Now, let's prove these solutions here using theorems.

- **Theorem 1: For any real numbers x and y, sin x = sin y implies x = nπ + (–1)ny, where n ∈ Z**

Proof: Consider the equation sin x = sin y and determine its general solution in trigonometry.

$$sin\,y = sin\,x$$

$$\Rightarrow sin\,x - sin\,y = 0$$

$$\Rightarrow 2cos(x+y)/2\,sin(x-y)/2 = 0$$

$$\Rightarrow cos(x+y)/2 = 0\,or\,sin(x-y)/2 = 0$$

Upon taking the common solution from both the conditions, we get:

$$x = n\pi + (-1)ny, where\,n \in Z$$

- Theorem 2: For any real numbers x and y, cos x = cos y implies x = 2nπ ± y, where n ∈ Z.

Proof: The general solution of cos x = cos y will be:

$$cos\,x - cos\,y = 0$$

$$2sin(x+y)/2\,sin(y-x)/2 = 0$$

$$sin(x+y)/2 = 0\,or\,sin(x-y)/2 = 0$$

$$(x+y)/2 = n\pi\,or\,(x-y)/2 = n\pi$$

On taking the common solution from both conditions, we get:

$$x = 2n\pi \pm y, where\,n \in Z$$

- Theorem 3: Prove that if x and y are not odd multiples of π/2, then tan x = tan y implies x = nπ + y, where n ∈ Z.

Proof: Likewise, when dealing with equations that contain trigonometric functions like tan x or others, we can employ trigonometric equation conversions to determine their solutions.

In other words, if $tan\,x = tan\,y$, then;

$$\frac{Sin\,x}{\cos x} = \frac{Sin\,y}{\cos y}$$

$$sin\,x\,cos\,y = sin\,y\,cos\,x = 0$$

$$sin\,x\,cos\,y - sin\,y\,cos\,x = 0$$

$sin(x - y) = 0$ [By trigonometric identity]

Hence, $x - y = n\pi \; or \; x = n\pi + y, where \; n \in Z.$

1.6.4. Circular Functions

Trigonometric functions, often referred to as Circular Functions, can be described as mathematical functions that connect angles within a triangle. In essence, these functions serve as a mathematical framework for understanding how the angles and sides of a triangle relate to each other. The core trigonometric functions include sine, cosine, tangent, cotangent, secant, and cosecant. A variety of trigonometric formulas and identities exist to establish these relationships and aid in calculating the angles of a triangle. (Bushell & Edmunds, 2012).

1.6.4.1. Six Circular Functions

Trigonometric functions are primarily categorized based on three fundamental angles: sine, cosine, and tangent.

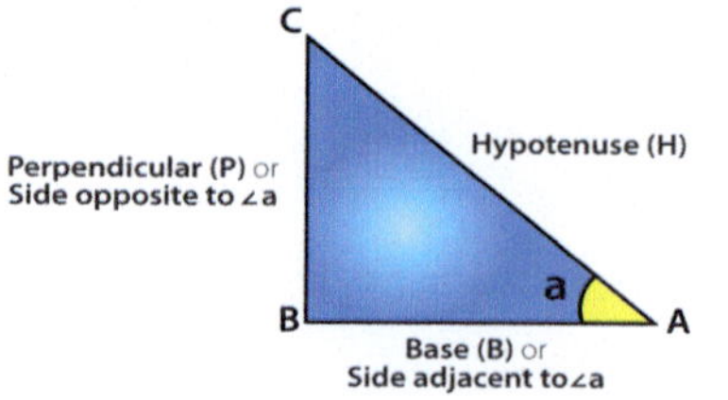

Figure 1.9. Trigonometric functions.

Source: Byjus, Creative Commons License.

The secondary functions, cotangent, secant, and cosecant, are derived from these fundamental functions but are less commonly used. Trigonometry is typically taught using right-angled triangles as a foundation (Stewart, 2022) (Figure 1.9).

a. Sine Function: Based on the above diagram, the value of the sine function will be determined using the following relation (Chen et al., 2003):

$$sin\, a = \frac{opposite}{hyoptenuse} = \frac{CB}{CA}$$

b. Cos Function: The cosine function for the above diagram can be derived as follows (Stolwijk et al., 1999):

$$cosa = \frac{adjacent}{hypotenuse} = \frac{AB}{CA}$$

c. Tan Function: Based on the figure above, the tangent function can be expressed as follows (Xie et al., 2018):

$$tan\, a = \frac{opposite}{adjacent} = \frac{CB}{BA}$$

Also, in terms of sine and cos, tan function can be represented as:

$$tana = sina\, / cosa$$

1.6.4.2. Secant, Cosecant, and Cotangent Functions

The reciprocals of sine, cosine, and tangent functions are cosecant (csc), secant (sec), and cotangent (cot). The formulas for each of these functions are provided as follows (Qureshi et al., 2019):

a. $$Sec\ a = 1/\left(cos\ a\right) = \frac{Hypotenuse}{Adjacent} = \frac{CA}{AB}$$

b. $$Cot\, a = 1/\left(tan\, a\right) = \frac{Adjacent}{Opposite} = \frac{BA}{CB}$$

c. $$Cosec\ a = \frac{1}{sin\ a} = \frac{Hypotenuse}{Opposite} = \frac{CA}{CB}$$

1.7. FORMULAS

This section will clarify the mathematical expressions provided in the table 1.2. These expressions relate to trigonometric ratios such as sine, cosine, tangent, cotangent, secant, and cosecant, and they are specifically designed for use in right-angled triangles (Kim, 2007).

Table 1.2. Formulas for Functions of Trigonometric Rations

The Formula for Angle θ	Reciprocal Identities
Cos θ = Adjacent side/Hypotenuse	cos $\theta = 1/sec\theta$
Sin θ = Opposite side/Hypotenuse	sin $\theta = 1/cosec\theta$
Tan θ = Opposite side/Adjacent	tan $\theta = 1/cot\theta$
Cot θ = Adjacent side / Opposite	cot $\theta = 1/tan\theta$
Cosec θ = Hypotenuse / Opposite	cosec $\theta = 1/sin\theta$
Sec θ = Hypotenuse / Adjacent side	sec $\theta = 1/cos\theta$

Source: Charles, Creative Commons License.

1.7.1. Identities

Below are the identities related to trig functions

1.7.1.1. Even and Odd Functions

The secant and cosine functions exhibit even symmetry, while the remaining functions display odd symmetry.

$$cos(-x) = cos\,x$$

$$sin(-x) = -sinx$$

$$tan(-x) = -tan\,x$$

$$sec(-x) = secx$$

$$cot(-x) = -cot\,x$$

$$csc(-x) = -csc\,x$$

1.7.1.2. Periodic Functions

Trigonometric functions belong to a category of functions that exhibit periodic behavior. In most trigonometric functions, the smallest periodic cycle occurs every 2π radians, except for the tangent and cotangent functions, which have a minimum repeating cycle of π radians. (Kämmerer et al., 2015).

$$cos(x+2n\pi)=cos\,x$$

$$sin(x+2n\pi)=sin\,x$$

$$tan(x+n\pi)=tan\,x$$

$$cot(x+n\pi)=cot\,x$$

$$sec(x+2n\pi)=sec\,x$$

$$csc(x+2n\pi)=csc\,x$$

where n is an integer.

1.7.1.3. Pythagorean Identities

When represented using trigonometric functions, the Pythagorean theorem is commonly referred to as the Pythagorean identity. There exist three primary identities (Zimba, 2009):

$$sin^2x+cos^2x=1$$

$$cosec^2x=1+cot^2x$$

$$1+tan^2x=sec^2x$$

1.7.1.4. Sum and Difference Functions

Following are the sum and difference functions:

1. $sin(x-y)=sin(x).cos(y)-cos(x).sin(y)$
2. $sin(x+y)=sin(x).cos(y)+cos(x).sin(y)$
3. $cos(x-y)=cosx.cosy+sinx.siny$

4. $cos(x+y) = cosx.cosy - sinx.siny$

5. $tan(x-y) = [tan(x) - tan(y)]/[1 + tan(x)tan(y)]$

6. $tan(x+y) = [tan(x) + tan(y)]/[1 - tan(x)tan(y)]$

1.7.2. Table and Graph

The table 1.3 shows the importance of the tables and graphs in trigonometry:

Table 1.3. Trigonometric Ratio Table for Six Trigonometric Functions

Angle Value / Trigonometric Functions	0°	30°	45°	60°	90°
Cosθ	1	$\sqrt{3}/2$	$1/\sqrt{2}$	1/2	0
Sinθ	0	½	$1/\sqrt{2}$	$\sqrt{3}/2$	1
Tanθ	0	$1/\sqrt{3}$	1	$\sqrt{3}$	∞
Cosecθ	∞	2	$\sqrt{2}$	$2/\sqrt{3}$	1
Cotθ	∞	$\sqrt{3}$	1	$1/\sqrt{3}$	0
Secθ	1	$2/\sqrt{3}$	$\sqrt{2}$	2	∞

Source: Charles, Creative Commons License.

Now, the formulas and numerical values associated with various angles for all trigonometric functions are known. To comprehend the variations over a given time interval, the graphical representations of the six trigonometric functions shall be examined (Abramowitz et al., 1988) (Figure 1.10) (Table 1.4).

Table 1.4. Domain & Range of each Function Graphed in XY Plane

Function	Definition	Domain	Range
Cosine Function	$y = cos\, x$	$x \epsilon R$	$-1 \leq cos\, x \leq 1$
Sine Function	$y = sinx$	$x \epsilon R$	$-1 \leq sin\, x \leq 1$
Tangent Function	$y = tanx$	$x \epsilon R, x \neq (2k+1)\pi / 2$	$-\infty < tan\, x < \infty$
Cotangent Function	$y = cot\, x$	$x \epsilon R, x \neq k\pi$	$-\infty < cot\, x < \infty$
Cosecant Function	$y = cscx$	$x \epsilon R, x \neq k$	$csc x \in (-\infty, -1) \cup (1, \infty)$
Secant Function	$y = sec\, x$	$x \epsilon R, x \neq (2k+1)\pi / 2$	$sec x \in (-\infty, -1) \cup (1, \infty)$

Source: Charles, Creative Commons License.

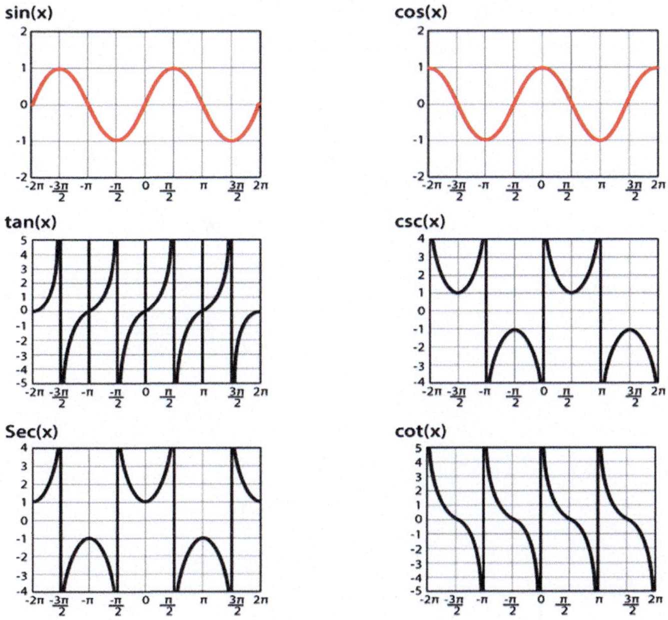

Figure 1.10. Graph for all functions based on their respective domain and range.

Source: Byjus, Creative Commons License.

Remember

Trigonometric concepts have been used to create brain-teasing puzzles and riddles that challenge mathematical minds.

1.8. UNIT CIRCLE

The unit circle is one of the basic geometric shapes in the field of trigonometry. The circle given below, positioned at the origin of a coordinate plane and possessing a radius of one unit, provides significant insights into the realm of angles, functions, and complicated interactions (Maknun et al., 2019).

The unit circle is not merely conventional but rather a geometric construct that exhibits the concept of unity in mathematics. By utilizing a straightforward mathematical equation, denoted as $x^2+y^2=1$, a trajectory of points is generated that maintains a consistent distance of one unit from the origin, which is located at coordinates (0, 0). Although the equation of the unit circle may appear simple, it has a vast array of mathematical connections (Figure 1.11).

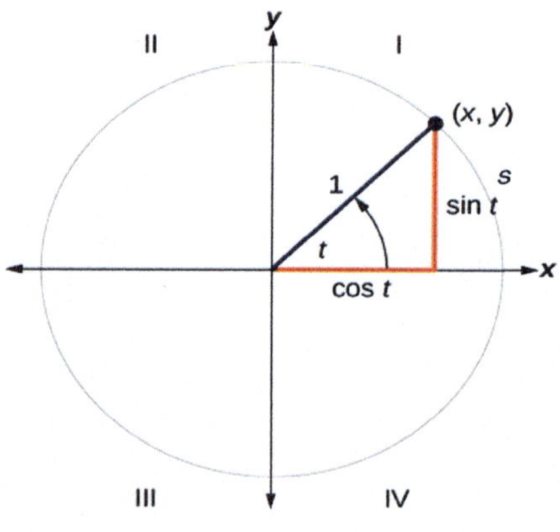

Figure 1.11. Unit circle.

Source: Mary Jane, Creative Commons License.

The unit circle is strongly connected to other geometric elements such as angles, arcs, coordinates, and triangles. Each angle, regardless of size, can correlate with a specific point on the unit circle. To achieve a comprehensive understanding of this concept, it is advisable to take into account an angle denoted as θ. When positioning the terminal side of angle θ on the unit circle, the coordinates (x, y) of the point of intersection corresponding to the values of sine and cosine of θ are:

$$x = cos\theta$$

$$y = sin\theta$$

The inherent connection between angles and coordinates establishes the unit circle as a potent instrument for computing trigonometric values.

Did You Know?

The unit circle is a powerful tool in trigonometry, where its radius is 1 unit. It's used extensively to define trigonometric functions and understand their properties.

The unit circle is a seamless connection between two distinct units of angular measurement, namely radians and degrees. By completing a single revolution along the circumference of a circle, an angular displacement of 2π radians or 360 degrees is covered. Utilizing this bi-angular capability allows for the seamless translation between the two angular frameworks, augmenting our comprehension of the various manifestations of angles (Bornstein, 2020).

The unit circle exhibits a harmonious interplay of trigonometric functions. Reexamine the three fundamental trigonometric functions, namely sine (sin), cosine (cos), and tangent (tan), and investigate their interplay within a unit circle.

The sine function, denoted as sin, is expressed as the y-coordinate of a point located on the unit circle (Altman & Kidron, 2016):

$$sin\theta = y$$

Cosine (cos) resonates through the x-coordinate:

$$cos\theta = x$$

Tangent (tan) takes its cue from the ratio of sine to cosine:

$$tan\theta = \frac{sin\theta}{cos\theta} = \frac{y}{x}$$

The relationships above, intricately intertwined within the structure of a unit circle, provide a visual manifestation of trigonometric functions applicable to all angles. The unit circle is divided into four quadrants, each characterized by unique signs of x and y coordinates. These quadrants provide insights into the behavior of trigonometric functions as they vary over different angles. The quadrantal angles at 0, 90, 180, and 270 degrees exhibit a direct alignment with the axes. This alignment facilitates the easy recognition of their corresponding sine and cosine values. In addition to these angles, the unit circle is a valuable tool for determining the signs and values of functions inside each quadrant (Martínez-Planell & Delgado, 2016).

Activity 1.1

Imagine you are an architect designing a bridge over a river. You need to calculate the angle of elevation to ensure the bridge's proper slope for safe transportation. How would you use trigonometric ratios to determine this angle, and what measurements or data would you need?

1.9. SUMMARY

This chapter titled "Introduction to Trigonometry" provides a comprehensive and enlightening entry point into the domain of angles, their interconnections, and the underlying mathematical structures. Trigonometry, an academic discipline with a long historical lineage and contemporary practical relevance, offers a systematic framework for deciphering the intricate geometric phenomena that influence the structure of our cosmos.

The voyage commences with a fundamental examination of angles, which serve as the foundational elements of trigonometry. The conversion from degrees to radians allows us to comprehend and quantify rotations, enabling a deeper comprehension of how angles permeate different facets of our existence.

Trigonometric ratios serve as fundamental tools for navigating the geometric properties of triangles. By utilizing the trigonometric functions of sine, cosine, and tangent, the intricate relationship between angles and sides can be elucidated.

The Pythagorean theorem is a crucial nexus, illustrating the profound symbiosis between geometry and algebra. This theorem demonstrates the interconnectedness of mathematical concepts by establishing a relationship between the lengths of sides in a right triangle.

The unit circle, a seemingly uncomplicated geometric entity, manifests as a remarkable feat of mathematical prowess. The concept of the unit circle connects angles and functions, providing a visual depiction that conveys significant insights into the complex interconnections among sine, cosine, and tangent.

As one navigates through the realm of trigonometric functions, their visual depictions elegantly adorn the mathematical landscape. Sine and cosine waves exhibit beautiful modulations, while tangent spikes provide insights into the exciting realm of oscillations.

Trigonometry's influence is readily seen in various domains, encompassing the prediction of wave behavior, the mapping of celestial trajectories, and the development of intricate engineering feats.

REFERENCES

1. Abramowitz, M., Stegun, I. A., & Romer, R. H., (1988). *Handbook of Mathematical Functions with Formulas, Graphs, and Mathematical Tables, 56(10), 958.*
2. Altman, R., & Kidron, I., (2016). Constructing knowledge about the trigonometric functions and their geometric meaning on the unit circle. *International Journal of Mathematical Education in Science and Technology, 47*(7), 1048–1060.
3. Arivina, A. N., & Jailani, J., (2020). Development of trigonometry learning kit with a STEM approach to improve problem solving skills and learning achievement. *Jurnal Riset Pendidikan Matematika, 7*(2), 178–194.
4. Bayeh, C., Bernard, M., & Moubayed, N., (2009). Introduction to the elliptical trigonometry. *WSEAS Transactions on Mathematics, 8*(9), 551–560.
5. Berggren, J. L., (2010). The joint journey of mathematics and astronomy in the ancient world. In: *AIP Conference Proceedings* (Vol. 1283, pp. 1–17). American Institute of Physics.
6. Bernard, M., Sumarna, A., Rolina, R., & Akbar, P., (2019). Development of high school student work sheets using VBA for Microsoft word trigonometry materials. In: *Journal of Physics: Conference Series* (Vol. 1315, pp. 12031). IOP Publishing.
7. Birman, G. S., & Nomizu, K., (1984). Trigonometry in Lorentzian geometry. *The American Mathematical Monthly, 91*(9), 543–549.
8. Blackett, N., & Tall, D. O., (1991). Gender and the versatile learning of trigonometry using computer software. In: *Proceedings of the 15th Conference of the International Group for the Psychology of Mathematics Education* (Vol. 1, pp. 144–151). Assisi: PME.
9. Bornstein, N., (2020). Teaching transformations of trigonometric functions with technology. *Journal of Interactive Media in Education, 1*(15), 1–9.
10. Bressoud, D. M., (2010). Historical reflections on teaching trigonometry. *The Mathematics Teacher, 104*(2), 106–112.
11. Brown, T. M., Casertano, S., Strader, J., Riess, A., VandenBerg, D. A., Soderblom, D. R., & Salinas, R., (2018). A high-precision trigonometric parallax to an ancient metal-poor globular cluster. *The Astrophysical Journal Letters, 856*(1), L6.

12. Bushell, P. J., & Edmunds, D. E., (2012). Remarks on generalized trigonometric functions. *The Rocky Mountain Journal of Mathematics, 6*(4), 25–57.
13. Chambers, P., (1999). Teaching Pythagoras' Theorem. *Mathematics in School, 28*(4), 22–24.
14. Chen, C. P., Zhao, J. W., & Qi, F., (2003). Three inequalities involving hyperbolically trigonometric functions. *RGMIA Research Report Collection, 6*(3), 359–364.
15. Demir, Ö., & Heck, A., (2013). A new learning trajectory for trigonometric functions. In: *Proceedings of the 11th International Conference on Technology in Mathematics Teaching* (Vol. 2, pp. 119–124).
16. Edmunds, D. E., Gurka, P., & Lang, J., (2012). Properties of generalized trigonometric functions. *Journal of Approximation Theory, 164*(1), 47–56.
17. Fahrudin, D., & Pramudya, I., (2019). Profile of students' errors in trigonometry equations. In: *Journal of Physics: Conference Series* (Vol. 1188, pp. 120–144). IOP Publishing.
18. Gel-fand, I. M., Saul, M., Gelfand, I. M., & Saul, M., (2001). *Trigonometry* (Vol. 2, pp. 1–20). Birkhäuser Boston.
19. Gerhana, M. T. C., Mardiyana, M., & Pramudya, I., (2017). The effectiveness of project-based learning in trigonometry. In: *Journal of Physics: Conference Series* (Vol. 895, pp. 120–127). IOP Publishing.
20. Gur, H., (2009). Trigonometry learning. *New Horizons in Education, 57*(1), 67–80.
21. Herrmann, D., (2023). Beginnings of trigonometry. In: *Ancient Mathematics: History of Mathematics in Ancient Greece and Hellenism* (Vol. 2, pp. 265–277). Berlin, Heidelberg: Springer Berlin Heidelberg.
22. Hirsch, C. R., Weinhold, M., & Nichols, C., (1991). Implementing the standards: Trigonometry today. *The Mathematics Teacher, 84*(2), 98–106.
23. Kämmerer, L., Potts, D., & Volkmer, T., (2015). Approximation of multivariate periodic functions by trigonometric polynomials based on rank-1 lattice sampling. *Journal of Complexity, 31*(4), 543–576.
24. Kendal, M., & Stacey, K., (1998). Teaching trigonometry. *Australian Mathematics Teacher, the, 54*(1), 34–39.
25. Kim, T., (2007). q-extension of the Euler formula and trigonometric

functions. *Russian Journal of Mathematical Physics, 14*(3), 275–278.

26. Kline, M., (1990). *Mathematical Thought from Ancient to Modern Times* (Vol. 2, pp. 87–90). Oxford University Press.
27. Knorr, W., (1985). Ancient versions of two trigonometric lemmas. *The Classical Quarterly, 35*(2), 362–391.
28. Maknun, C. L., Rosjanuardi, R., & Jupri, A., (2019). From ratios of right triangle to unit circle: An introduction to trigonometric functions. In: *Journal of Physics: Conference Series* (Vol. 1157, pp. 221–224). IOP Publishing.
29. Martínez-Planell, R., & Delgado, A. C., (2016). The unit circle approach to the construction of the sine and cosine functions and their inverses: An application of APOS theory. *The Journal of Mathematical Behavior, 43*, 111–133.
30. Moore, K. C., & LaForest, K. R., (2014). The circle approach to trigonometry. *The Mathematics Teacher, 107*(8), 616–623.
31. Moore, K. C., LaForest, K., & Kim, H. J., (2012). The unit circle and unit conversions. In: *Proceedings of the Fifteenth Annual Conference on Research in Undergraduate Mathematics Education* (pp. 16–31). Portland State University Portland.
32. Ngu, B. H., & Phan, H. P., (2020). Learning to solve trigonometry problems that involve algebraic transformation skills via learning by analogy and learning by comparison. *Frontiers in Psychology, 11*, 558773.
33. Otero, D., (2022). *The Trigonometric Functions Through Their Origins: Regiomontanus and the Beginnings of Modern Trigonometry, 74*(4), 852–856.
34. Qureshi, M. I., Porwal, S., Ahamad, D., & Quraishi, K. A., (2019). Successive differentiations of tangent, cotangent, secant, cosecant functions and related hyperbolic functions (A hypergeometric approach). *International Journal of Mathematics Trends and Technology, 65*(7), 352–367.
35. Rosenberg, H., (1958). The changing concept of trigonometry as a school subject. *The Mathematics Teacher, 51*(4), 246–252.
36. Saville, D. J., Wood, G. R., Saville, D. J., & Wood, G. R., (1991). The geometric tool kit. *Statistical Methods: The Geometric Approach*, 10–38.

37. Stewart, S. M., (2022). A look back at a long-forgotten trigonometric function: The versine function and its inverse. *The Mathematical Gazette, 106*(565), 84–94.
38. Stolwijk, A. M., Straatman, H. M. P. M., & Zielhuis, G. A., (1999). Studying seasonality by using sine and cosine functions in regression analysis. *Journal of Epidemiology & Community Health, 53*(4), 235–238.
39. Strang, G., (1986). *Introduction to Applied Mathematics* (Vol. 1, pp. 1–200). Wellesley-Cambridge Press.
40. Tuna, A., & Kacar, A., (2013). The effect of 5E learning cycle model in teaching trigonometry on students' academic achievement and the permanence of their knowledge. *International Journal on New Trends in Education and Their Implications, 4*(1), 73–87.
41. Ungar, A. A., (2001). Hyperbolic trigonometry and its application in the Poincaré ball model of hyperbolic geometry. *Computers & Mathematics with Applications, 41*(1–2), 135–147.
42. Urrutia, F. Z., Loyola, C. C., & Marín, M. H., (2019). A tangible user interface to facilitate learning of trigonometry. *International Journal of Emerging Technologies in Learning (IJET), 14*(23), 152–164.
43. Webber, W. P., (1931). Introduction to trigonometry from the historical view point. *Mathematics News Letter, 10*(7), 14–17.
44. Wijaya, T. T., Ying, Z., & Purnama, A., (2020). Using Hawgent dynamic mathematic software in teaching trigonometry. *International Journal of Emerging Technologies in Learning (IJET), 15*(10), 215–222.
45. Xie, G. S., Jin, X. B., Yang, C., Pu, J., & Mo, Z., (2018). Graphical representation and similarity analysis of DNA sequences based on trigonometric functions. *Acta Biotheoretica, 66*, 113–133.
46. Yang, D. C., & Sianturi, I. A., (2017). An analysis of Singaporean versus Indonesian textbooks based on trigonometry content. *Eurasia Journal of Mathematics, Science and Technology Education, 13*(7), 3829–3848.
47. Zengin, Y., Furkan, H., & Kutluca, T., (2012). The effect of dynamic mathematics software GeoGebra on student achievement in teaching of trigonometry. *Procedia-Social and Behavioral Sciences, 31*, 183–187.
48. Zimba, J., (2009). On the possibility of trigonometric proofs of the Pythagorean theorem. In: *Forum Geometricorum* (Vol. 9, pp. 1–4).

CHAPTER 2

The Tangent

Contents

2.1. INTRODUCTION

The tangent function (tan) and tangent lines are fundamental concepts in mathematics with wide-ranging applications in geometry, trigonometry, calculus, and various scientific disciplines. The tangent function provides a fundamental framework for understanding angles and right triangles, as it relates the angle (θ) to the ratio of the perpendicular (opposite) to the base (adjacent) in a right triangle: $\left(\tan(\theta)=\frac{\text{opposite}}{\text{adjacent}}\right)$. Beyond trigonometry, it is an essential tool in calculus, where it aids in the determination of instantaneous rates of change through differentiation.

Tangent lines, on the other hand, are a geometric concept related to curves and functions. A tangent line to a curve at a specific point is a straight line that touches the curve at that point without crossing it. It represents the instantaneous slope or rate of change of the curve at that particular location. In mathematical terms, the slope of a tangent line at a point is equivalent to the derivative of the function representing the curve at that point. This concept is fundamental in calculus, where it enables the analysis of functions' behavior, identification of critical points, and optimization. Tangent lines offer local linear approximations, simplifying complex functions into simpler linear models near particular points, a technique widely employed in mathematical modeling and problem-solving.

2.2. DEFINITION OF THE TANGENT FUNCTION (TAN Θ)

In practical terms, the tangent function (tan θ) is a mathematical tool that helps us understand the slope or inclination of a line or surface. It's often used when dealing with angles, inclines, or rates of change. Here's how it works:

- **Slope of a Line:** The slope of the hill can be measured as the ratio of how much you go up (rise) compared to how much you move horizontally (run). This slope is akin to the tangent of the angle of inclination.
- **Inclined Surfaces:** When you encounter an inclined surface, like a ramp, the steepness of the ramp is related to the tangent of the angle of inclination. A steeper ramp has a higher tangent value.
- **Rates of Change:** In calculus, the tangent function is used to express the rate of change of a function at a particular point. The

tangent line to the curve at that point represents how the function changes locally.

Mathematically, for a given angle θ:

$$\tan(\theta) = \frac{\text{Height}}{\text{Base}}$$

Height represents how much you go up (rise).

The base represents how far you move horizontally (run).

In essence, the tangent function quantifies the steepness or slope of an angle or a curve. It is used in fields such as construction (for determining slopes), physics (for analyzing inclines and forces), and calculus (finding rates of change).

For example, if you're building a wheelchair ramp, you'd want to calculate the tangent of the angle to ensure it's not too steep for easy access. In physics, when studying a hill's incline, the tangent function helps to understand how gravity affects objects moving on the slope.

2.3. THE ROLE OF TANGENT IN RIGHT TRIANGLES

The role of the tangent function in right triangles is to establish a fundamental trigonometric relationship between the lengths of the sides of the triangle and the measure of one of its angles. This relationship is particularly useful when we know the lengths of two sides of a right triangle and need to find the measure of an angle or when we know an angle and one side and need to find another side's length (Figure 2.1).

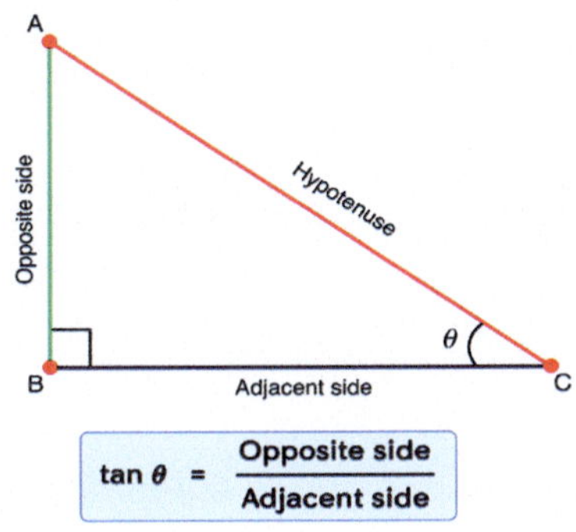

Figure 2.1. Tangent in a triangle.

Source: Math Monks, Creative Commons License.

Mathematically, for a right triangle with an acute angle θ:

$$\tan(\theta) = \frac{\text{Length of the opposite side}}{\text{Length of the adjacent side}}$$

The tangent of angle θ (tan θ) is defined as the ratio of the length of the side opposite angle θ (the "opposite" side) to the length of the side adjacent to angle θ (the "adjacent" side).

Given the lengths of the opposite and adjacent sides, you can find the measure of angle θ using the inverse tangent function (arctan or $\tan^{-1}$). The formula is:

$$\theta = \tan^{-1}\left(\frac{\text{Length of the opposite side}}{\text{Length of the adjacent side}}\right)$$

Conversely, if you know the measure of angle θ and the length of either the opposite or adjacent side, you can find the length of the other side using the tangent function:

$$\text{Length of the opposite side} = \text{Length of the adjacent side} \cdot \tan(\theta)$$

or

$$\text{Length of the adjacent side} = \frac{\text{Length of the opposite side}}{\tan(\theta)}$$

For instance, in surveying, if you measure the height of a building (opposite side) and your distance from the building (adjacent side), you can use the tangent function to find the angle of elevation to the top of the building. Conversely, if you know the angle of elevation and your distance, you can calculate the building's height.

2.4. BASIC PROPERTIES OF THE TANGENT FUNCTION

Here are some basic properties of the Tangent Function:

2.4.1. Periodicity

The tangent function is periodic with a period of π radians (180 degrees).

Mathematically, it repeats its values every π radians:

$$\tan(\theta) = \tan(\theta + \pi) = \tan(\theta + 180°)$$

Example:

$$\tan(30°) = \tan(30° + 180°) = \tan(210°)$$

This is because they all correspond to the same position on the unit circle.

2.4.2. Asymptotes

The tangent function has vertical asymptotes at odd multiples of π/2 radians (90 degrees).

At these points, the tangent function approaches positive or negative infinity, depending on the direction of the approach.

Mathematically:

$$\left[\tan(\theta) \rightarrow +\infty \text{ as } \theta \rightarrow (2n+1)\frac{\pi}{2}\right]$$

$$\left[\tan(\theta) \rightarrow -\infty \text{ as } \theta \rightarrow (2n-1)\frac{\pi}{2}\right]$$

Example:

i. As (θ) approaches $(90°)(\pi / 2 \ \textit{radians}) \ \textit{from the left side}, \tan(\theta)$ approaches positive infinity.

ii. As (θ) approaches $(270°)$ $(3\pi / 2$ radians) from the left side, tan (θ) approaches negative infinity.

2.4.3. Range

The range of the tangent function is all real numbers.

Mathematically:

$$\textit{Range of } \tan(\theta) = (-\infty, +\infty)$$

2.4.4. Graphical Behavior

The graph of the tangent function exhibits a periodic, wave-like pattern with asymptotes at regular intervals (Figure 2.2).

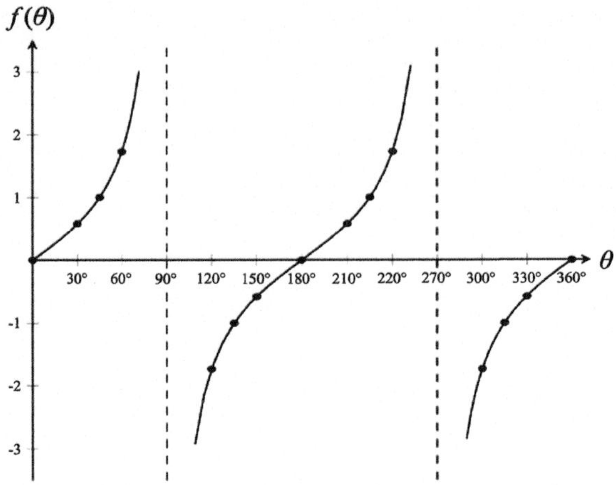

Figure 2.2. Graph of a tangent function.

Source: Nigerian Scholars, Creative Commons License.

It approaches positive or negative infinity at the vertical asymptotes.

- **Reciprocal Identity:** The tangent function has a reciprocal identity as follows:

$\left(\tan(\theta) = \dfrac{1}{\cot(\theta)}\right)$, here cot (θ) is the cotangent function.

- Example:

$$\text{If} \tan(\theta) = 2, \text{then} \left(\cot(\theta) = \frac{1}{2}\right).$$

Understanding these properties of the tangent function is essential when working with trigonometric equations, graphs, and real-world applications. It allows us to analyze its behavior, identify key points, and make accurate calculations in various contexts.

2.5. GRAPHICAL REPRESENTATION OF THE TANGENT FUNCTION

The tangent function tan (θ) is known for its periodic, wave-like graph. Here are some key characteristics of the graph of tan (θ) function:

- **Periodicity:** The tangent function has a period of π radians (180 degrees). This means it repeats its values every π radians.

The graph extends infinitely in both the positive and negative directions along the x-axis.

- **Vertical Asymptotes:** The tangent function has vertical asymptotes at odd multiples of π/2 radians (90 degrees). At these points, the function approaches positive or negative infinity, depending on the direction of the approach.
- **Amplitude and Range:** The tangent function's amplitude is not limited like sine and cosine; it can take on all real values. Its range is all real numbers, so it extends infinitely both upwards and downwards.
- **Symmetry:** The tangent function has no symmetry with respect to the y-axis or origin.

Here is a graphical representation of the tangent function $y = \tan(\theta)$ (Figure 2.3):

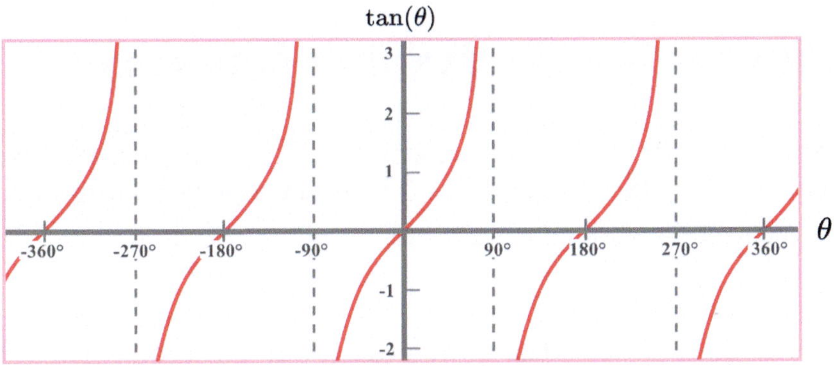

Figure 2.3. Graphical representation of the tangent function $y = \tan(\theta)$

Source: *Third Space Learning, Creative Commons License.*

In this graph, you can observe the following:

- The graph repeats itself every π radians.
- Vertical asymptotes occur at odd multiples of π/2 radians (90 degrees), which are the points where the function approaches positive or negative infinity.
- The function approaches zero as θ approaches multiples of π radians (180 degrees).

- The graph is continuous, but it exhibits abrupt changes in slope at the vertical asymptotes.

Example: Let's find the values of: $(\tan(0)), (\tan(\pi/4)), \text{and} (\tan(\pi/2))$

$$(\tan(0) = 0)$$

$$(\tan(\pi/4) = 1)$$

$$(\tan(\pi/2) = \infty)$$

These values correspond to points on the graph where the tangent function intersects the x-axis (θ = 0), where it reaches its maximum (θ = π/4), and where it has a vertical asymptote (θ = π/2).

Understanding the graphical representation of the tangent function is crucial when analyzing trigonometric equations and solving real-world problems involving angles and inclines.

2.6. SOLVING SIMPLE TANGENT EQUATIONS

Simple tangent equations involve finding the value of the angle (θ) when given the value of the tangent function (tan θ). We can use the inverse tangent function, often denoted as $\tan^{-1}$ or arctan, to solve such equations.

If you have an equation in the form:

$$\tan(\theta) = \text{value}$$

You can solve for θ by taking the arctan (inverse tangent) of both sides:

$$\theta = \tan^{-1}(\text{value})$$

Example: Solve for θ in the equation: $\tan(\theta) = 1$.

Solution: Using the inverse tangent function, we have:

$$\theta = \tan^{-1}(1)$$

Now, calculate the value of (θ) by finding the angle whose tangent is 1.

This occurs at $\left(\theta = \frac{\pi}{4}\right)$ radians or $(\theta = 45°)$.

So,

$(\theta = \frac{\pi}{4}$ radians or $(\theta = 45°)$.

Example: Solve for θ in the equation: $\left(\tan(\theta) = -\sqrt{3}\right)$.

Solution: Again, using the inverse tangent function, we have:

$\theta = \tan^{-1}\left(-\sqrt{3}\right)$

Calculate the value of (θ) by finding the angle whose tangent is -√3.

This occurs at $\left(\theta = -\frac{ð}{3}\right)$ radians or $(\theta = -60°)$.

So,

$(\theta = -\frac{\pi}{3}$ radians or $(\theta = -60°)$.

Note: It's important to remember that the tangent function is periodic, and angles that differ by multiples of π (180°) have the same tangent value. Therefore, when solving tangent equations, there may be multiple solutions within a period (e.g., 0 to π or 0° to 180°).

These examples demonstrate how to solve simple tangent equations to find the angle θ when given a specific tangent value. The inverse tangent function, often available on calculators as $\tan^{-1}$ or arctan, is a valuable tool for these calculations.

2.7. PERIODICITY AND BEHAVIOR OF TANGENT

The periodicity of tan θ is best illustrated when graphing the function, revealing its wave-like pattern, asymptotes, and how it approaches positive or negative infinity at these critical points. Understanding this periodic behavior is fundamental in trigonometry and calculus when dealing with angles, slopes, and rates of change.

For any angle θ, the tangent function at θ is equal to the tangent function at θ plus any multiple of the period (π radians):

$$\tan(\theta) = \tan(\theta + n\pi)$$

Here, 'n' is an integer representing the number of periods that have been added.

Example: Consider the tangent function at θ = π/4 radians (45 degrees).

$$\tan\left(\frac{\pi}{4}\right) = \tan\left(\frac{\pi}{4} + \pi\right) = \tan\left(\frac{5\pi}{4}\right)$$

This demonstrates that the tangent function's value at π/4 is the same as its value at 5π/4, which is one period (π radians) ahead.

Graphically, the periodicity of the tangent function is evident when you plot its graph. The graph repeats its pattern every π radians, creating a wave-like appearance.

Understanding this periodic behavior is crucial in trigonometry, as it allows us to predict the values of the tangent function at various angles. It simplifies calculations involving angles and slopes and is a key element in trigonometric identities and equations.

2.8. AMPLITUDE AND RANGE OF THE TANGENT FUNCTION

Unlike sine and cosine functions, the tangent function does not have a defined amplitude. The amplitude of a function represents its maximum absolute value, and for the tangent function, this value is not bounded. Therefore, the tangent function can take on all real values.

The range of the tangent function is all real numbers, which means it can take on any real value, both positive and negative, and even zero.

Mathematically, the range of the tangent function is expressed as:

$$\text{Range of } \tan(\theta) = (-\infty, +\infty)$$

Example: Consider the tangent function evaluated at various angles θ:

$$(\tan(0) = 0)$$

$$\left(\tan\left(\frac{\pi}{4}\right) = 1\right)$$

$$\left(\tan\left(\frac{\pi}{2}\right) = \infty\right)(\textit{vertical asymptote})$$

$$\left(\tan\left(\frac{3\pi}{4}\right) = -1\right)$$

$$\left(\tan\left(3\pi/2\right) = -\infty\right)$$

These examples demonstrate that the tangent function can take on a wide range of values, from zero to positive and negative infinity. There is no upper or lower limit to the values it can attain.

Understanding the range of the tangent function is essential when working with trigonometric equations and graphs, as it helps identify the possible values that the function can produce, which are unbounded in both the positive and negative directions.

2.9. ASYMPTOTES AND POINTS OF DISCONTINUITY

For any integer 'n,' the vertical asymptotes of the tangent function occur at:

$$\theta = (2n+1)\frac{\pi}{2}$$

As θ approaches these values, the tangent function behaves as follows:

As (θ) approaches $(90°)(\pi/2\ radians)\ \textit{from the left side},\ \tan(\theta)$ approaches positive infinity.

As (θ) approaches $(270°)\,(3\pi/2$ radians) from the left side, $\tan(\theta)$ approaches negative infinity.

Points of Discontinuity: The tangent function has points of discontinuity at values of θ where the tangent function is undefined. The primary points of discontinuity correspond to the vertical asymptotes (odd multiples of π/2), where the function is undefined.

Remember

Tangent functions have asymptotes and points of discontinuity, which can affect the behavior of the function.

Example: Consider the tangent function at several key points:

$$\left(\tan\left(\frac{\pi}{4}\right) = 1\right)$$

$\tan\left(\frac{\pi}{2}\right)$ is undefined (vertical asymptote)

$$\tan\left(\frac{3\pi}{4}\right) = -1$$

$$\tan(\pi) = 0$$

$$\tan\left(\frac{5\pi}{4}\right) = 1$$

$\tan\left(\frac{3\pi}{2}\right)$ is undefined (vertical asymptote)

These examples illustrate that the tangent function is continuous except at the odd multiples of π/2, where it has vertical asymptotes and points of discontinuity.

2.10. TRANSFORMATIONS OF THE TANGENT FUNCTION

The tangent function can be transformed by applying various operations to the argument (θ) and the result (tan θ). These transformations include translations, reflections, stretches, and compressions. They alter the shape, period, and position of the tangent graph. Here are some common transformations:

- **Vertical Translation (Shift):** Adding or subtracting a constant (c) to the result (tanθ), shifts the entire graph vertically.

  Example: $(y = \tan(\theta) + 2)$ shifts the graph of $(y = \tan(\theta))$ upward by 2 units.
- **Horizontal Stretch or Compression:** Multiplying the argument θ by a constant (a) stretches or compresses the graph horizontally.

  Example: $(y = \tan(2\theta))$ compresses the graph horizontally by a factor of 2.
- **Phase Shift (Horizontal Translation):** Adding or subtracting a constant (b) inside the argument θ shifts the graph horizontally.

  Example: $\left(y = \tan\left(\theta - \frac{ð}{4}\right)\right)$ shifts the graph to the right by

$\left(\frac{\pi}{4}\right)$ radians.

- **Vertical Stretch or Compression:** Multiplying the result (tan θ) by a constant (k) stretches or compresses the graph vertically.

  Example: $(y = 2\tan(\theta))$ stretches the graph vertically by a factor of 2.

- **Reflections:** Reflecting the graph across the x-axis or y-axis results in mirror images.

  Example: $y = -\tan(\theta)$ reflects the graph across the x-axis.

  Example: Original Function: $(y = \tan(\theta))$

Transformed Function: $\left(y = -2\tan\left(2\theta - \frac{\pi}{3}\right) + 1\right)$

In this example, we have applied multiple transformations:

a. Vertical Stretch: The graph is stretched vertically by a factor of 2.

b. Horizontal Stretch: The graph is compressed horizontally by a factor of 2.

c. Phase Shift: The graph is shifted to the right by $\left(\frac{\pi}{3}\right)$ radians.

d. Vertical Translation: The entire graph is shifted upward by 1 unit.

These transformations allow us to manipulate the tangent function to fit specific needs in various mathematical and real-world contexts, such as physics, engineering, and signal processing.

2.11. APPLICATIONS OF THE TANGENT FUNCTION

This section will explore some key applications of the tangent function (tan θ) in trigonometry are abundant, especially in fields like engineering, physics, and geometry.

2.11.1. Slope and Inclination

In geometry and engineering, the tangent function is used to determine the slope or inclination of lines and surfaces. Given an inclined object, the tangent of the angle of inclination provides the slope. For example, a road with an angle of 5 degrees has a slope of tan (5)' which is approximately 0.0875 (Figure 2.4).

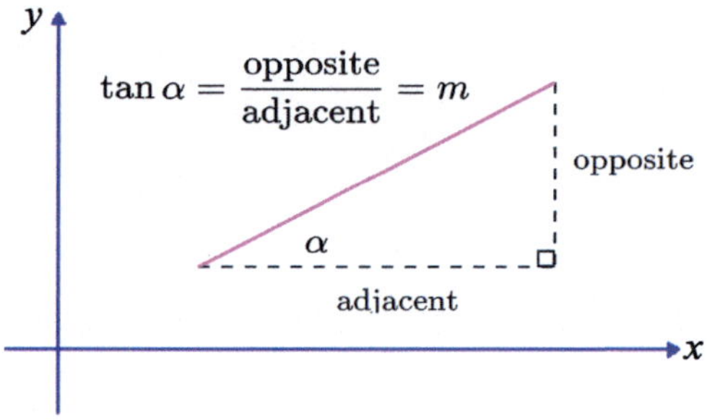

Figure 2.4. Slop of inclination using tangent.

Source: Interactive Mathematics, Creative Commons License.

2.11.2. Architecture and Construction

Architects and engineers use the tangent function to calculate the slope of roofs, ramps, and staircases, ensuring safe and functional designs. For instance, when designing a wheelchair ramp, the angle of the ramp's slope can be determined using the tangent function to comply with accessibility standards.

2.11.3. Trigonometric Surveying

In surveying, the tangent function helps measure angles accurately. Surveyors use theodolites (instrument) and total stations to calculate angles of elevation and depression.

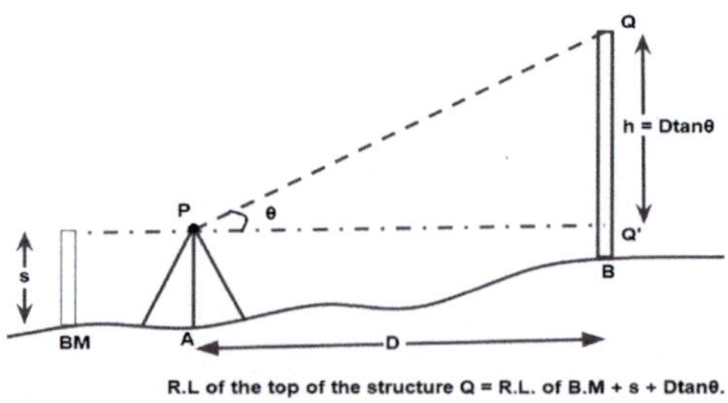

Figure 2.5. Trigonometric leveling using tangent function.

Source: ESE Notes, Creative Commons License.

The tangent formula is instrumental in finding distances, heights, and positions based on angular measurements. For instance, in triangulation, the tangent formula plays a crucial role in determining distances to inaccessible objects (Figure 2.5).

2.11.4. Physics and Mechanics

In physics, the tangent function is essential for analyzing the angles of inclined planes, slopes, and projectile motion. When studying inclined planes, the tangent function helps determine forces acting along and perpendicular to the slope. In projectile motion, the angle at which an object is launched affects its trajectory and range. The tangent function helps find the launch angle for desired outcomes.

Did You Know?

Tangent lines play a significant role in local linear approximations of trigonometric curves.

2.11.5. Trigonometric Identities

The tangent function is a fundamental component of trigonometric identities. For instance, the Pythagorean identity for tangent relates it to the sine and cosine functions: $\left(\tan^2(\theta) = \sec^2(\theta) - 1\right)$. These identities are used extensively in trigonometric simplifications, proofs, and calculations.

2.11.6. Electrical Engineering

In electrical engineering, the tangent function is employed to analyze phase differences and impedance in alternating current (AC) circuits. Engineers use the tangent of the phase angle in impedance calculations to determine the relationship between voltage and current in circuits.

2.11.7. Computer Graphics and Animation

Tangent functions are used to create smooth and realistic animations in computer graphics. They control the interpolation of keyframes, ensuring objects move smoothly along paths.

2.12. TANGENT LINES TO TRIGONOMETRIC CURVES

Tangent lines to trigonometric curves are straight lines that touch or intersect these curves at a single point without crossing or intersecting them elsewhere. These tangent lines provide valuable information about the behavior of trigonometric functions at specific points and are instrumental in understanding rates of change and local linear approximations (Figure 2.6).

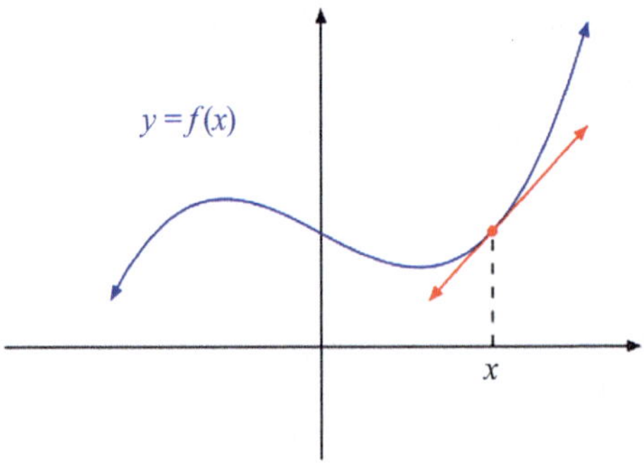

Figure 2.6. Illustration of a tangent line.

Source: EMORY, Creative Commons License.

Given a trigonometric function, such as $(y=\sin(\theta))or(y=\cos(\theta))$, a tangent line can be drawn to the curve at a particular angle $(\theta=a)$. This tangent line is characterized by its slope and equation.

1. Slope of the Tangent Line:

The slope m of the tangent line at an angle $(\theta=a)$ is equal to the derivative of the trigonometric function at that point. In mathematical terms:

$$m=\frac{dy}{d\theta}\Big|_{\theta=a}$$

For example, if you want to find the slope of the tangent line to $y=\sin(\theta)at\,\theta=\frac{\pi}{4}$, you calculate $\left(\frac{dy}{d\theta}\right)for\left(\theta=\frac{\pi}{4}\right)$.

2. Equation of the Tangent Line:

The equation of the tangent line at an angle $(\theta = a)$ is given by the point-slope form:

$$y - y_1 = m(x - x_1)$$

where (x_1, y_1) is the point of tangency (i.e., the point where the tangent line touches the curve).

Substituting the slope (m) and the angle $(\theta = a)$ into the equation, you obtain the specific equation of the tangent line.

Example: Consider the trigonometric function $(y = \sin(\theta))$. To find the equation of the tangent line at $\left(\theta = \frac{\pi}{4}\right)$, we follow these steps:

1. Calculate the slope of the tangent line:

$$\frac{dy}{d\theta} = \frac{d}{d\theta}(sin(\theta)) = \cos(\theta)$$

At $\left(\theta = \frac{\pi}{4}\right), (\cos\left(\frac{\pi}{4}\right) = \frac{1}{\sqrt{2}}.$

2. Use the point-slope form to find the equation of the tangent line:

Let's assume that the point of tangency is $\left(\frac{\pi}{4}, \sin\frac{\pi}{4}\right), which is \left(\frac{\pi}{4}, \frac{1}{\sqrt{2}}\right).$

The equation of the tangent line is:

$$y - \frac{1}{\sqrt{2}} = \frac{1}{\sqrt{2}}\left(\theta - \frac{\pi}{4}\right)$$

This equation represents the tangent line to the curve $(y = \sin(\theta))$ at $\left(\theta = \frac{\pi}{4}\right).$

Tangent lines to trigonometric curves are valuable for approximating the behavior of these curves locally, understanding rates of change, and making precise calculations in various applications, such as physics and engineering.

2.12.1. Tangent Lines in Trigonometry

In trigonometry, tangent lines play a crucial role in understanding the behavior of trigonometric functions, particularly sine and cosine, at specific points on their graphs.

Let's consider the sine function $(y=\sin(\theta))$ as an example. At any angle $(\theta=a)$ on the unit circle, we can draw a tangent line to the curve $(y=\sin(\theta))$ at that point. This tangent line is characterized by two key properties:

1. **Point of Tangency:** The point where the tangent line touches the curve $(y=\sin(\theta))$ is the point of tangency. This point is denoted as $(a,\sin(a))$.

2. **The slope of the Tangent Line:** The slope (m) of the tangent line at this point is equal to the derivative of the sine function at an angle $(\theta=a)$. In mathematical terms:

$$\left[m=\frac{dy}{d\theta}\Big|_{\theta=a}=\cos(a)\right]$$

Here, $\left(\frac{dy}{d\theta}\right)$ represents the derivative of $(y=\sin(\theta))$.

Example: Let's find the equation of the tangent line to the curve $(y=\sin(\theta))$ at the angle $\left(\theta=\frac{\pi}{4}\right)$:

1. Calculate the slope of the tangent line at $\left(\theta=\frac{\pi}{4}\right)$:

$$=\cos\left(-\right)=\frac{}{\sqrt{}}$$

2. Use the point-slope form of the equation of a line:

$$y-\sin\left(\frac{\pi}{4}\right)=\frac{1}{\sqrt{2}}\left(\theta-\frac{\pi}{4}\right)$$

This equation represents the tangent line to the sine curve at $\left(\theta=\frac{\pi}{4}\right)$.

Tangent lines in trigonometry are vital for understanding local behavior, making approximations, and calculating instantaneous rates of change. They are extensively used in physics, engineering, and calculus when analyzing the properties of trigonometric functions at specific points.

2.12.2. Finding Tangent Lines to Trigonometric Curves

In trigonometry, it is often necessary to find the equation of a tangent line to a trigonometric curve at a particular point. This process involves determining the slope of the tangent line at the given point and then using that slope to write the equation of the tangent line.

Let's consider the curve $(y=\sin(\theta))$ as our trigonometric curve and find the equation of the tangent line at a specific point, say, $\left(\theta=\frac{\pi}{3}\right)$.

1. Calculate the Slope m of the Tangent Line:

The slope of the tangent line at a point $(a,\sin(a))$ on the curve $(y=\sin(\theta))$ is given by the derivative of the sine function evaluated at $(\theta=a)$:

$$m=\frac{dy}{d\theta}\Big|_{\theta=a}=\cos(a)$$

In our case, at $\left(\theta=\frac{\pi}{3}\right)$, *the slope* (m) *is* $\left(\cos\left(\frac{\pi}{3}\right)=\frac{1}{2}\right)$.

2. Use the Point-Slope Form for the equation of a line:

The equation of the tangent line is expressed in point-slope form:

$$y-\sin(a)=m(\theta-a)$$

3. Substitute the Values

We know that $\left(a=\frac{\pi}{3}\right)$, $\left(\sin(a)=\sin\left(\frac{\pi}{3}\right)=\frac{\sqrt{3}}{2}\right)$, *and* $\left(m=\frac{1}{2}\right)$.

Substituting these values into the point-slope form, we get the equation of the tangent line:

$$y-\frac{\sqrt{3}}{2}=\frac{1}{2}\left(\theta-\frac{\pi}{3}\right)$$

Example: Let's find the equation of the tangent line to the curve $(y=\sin(\theta))at\left(\theta=\frac{\pi}{3}\right)$:

1. Calculate the slope:

$$m=\cos\left(\frac{\pi}{3}\right)=\frac{1}{2}$$

2. Use the point-slope form:

$$y-\frac{\sqrt{3}}{2}=\frac{1}{2}\left(\theta-\frac{\pi}{3}\right)$$

This equation represents the tangent line to the sine curve at $\left(\theta=\frac{\pi}{3}\right)$.

2.12.3. Using Derivatives to Determine Slopes and Equations of Tangent Lines

In calculus, derivatives play a crucial role in determining the slopes and equations of tangent lines to curves, including trigonometric curves. Here, we'll use the sine curve $(\mathrm{y}=\sin(\theta))$ as an example to illustrate how derivatives are employed to find the slopes and equations of tangent lines.

Given a curve, we want to find the equation of the tangent line at a specific angle (θ = a).

1. Calculate the Derivative $(dy/d\theta)$ of the Curve:

   The derivative of $y=\sin(\theta)$ with respect to (θ) is $\frac{dy}{d\theta}=\cos(\theta)$. This derivative represents the rate of change of (y) with respect to (θ), which is the slope of the curve at any angle (θ).

2. Evaluate the Derivative at the Given Angle:

To find the slope of the tangent line at $(\theta=a)$, we evaluate the derivative at that angle:

$$m=\frac{dy}{d\theta}\Big|_{\theta=a}=\cos(a)$$

3. Use the Point-Slope Form for the Equation of a Line:
The equation of the tangent line in point-slope form is:

$$y-y_1=m(\theta-a)$$

where $(a,\sin(a))$ is the point of tangency on the sine curve.

4. Substitute Values to Find the Equation:

Substitute the values of $(a),\sin(a)$, and (m) into the point-slope form to obtain the equation of the tangent line.

Example: Let's find the equation of the tangent line to the sine curve $(y=\sin(\theta))at\left(\theta=\frac{\pi}{3}\right)$:

1. Calculate the derivative:

$$\frac{dy}{d\theta}=\cos(\theta)$$

2. Evaluate the derivative at $\left(\theta=\frac{\pi}{3}\right)$:

$$m=\cos\left(\frac{\pi}{3}\right)=\frac{1}{2}$$

3. Use the point-slope form:

$$y-\sin\left(\frac{\pi}{3}\right)=\frac{1}{2}\left(\theta-\frac{\pi}{3}\right)$$

This equation represents the tangent line to the sine curve at $\left(\theta=\frac{\pi}{3}\right)$.

Using derivatives to determine slopes and equations of tangent lines is a fundamental concept in calculus. It allows us to understand the local behavior of functions and analyze rates of change in various mathematical and scientific applications.

2.12.4. Tangent Lines and Local Linear Approximations

Tangent lines are used to make local linear approximations of functions at specific points. This process involves finding the equation of the tangent line to a curve at a given point and using it to estimate the behavior of the function near that point. Let's illustrate this concept with a mathematical example.

Consider the curve $\left(y=\sqrt{x}\right),$ and we want to make a local linear approximation of the function near the point $\left(a,\sqrt{a}\right)$. Here's how we do it:

1. Calculate the Derivative (dy/dx) of the Curve:

   The derivative of $\left(y=\sqrt{x}\right)$ with respect to $(x)is\left(\frac{dy}{dx}=\frac{1}{2\sqrt{x}}\right)$.
2. Evaluate the Derivative at the Given Point:

   To find the slope of the tangent line at $(x=a)$, we evaluate the derivative at that point:

   $$\left[m=\frac{dy}{dx}\bigg|_{x=a}=\frac{1}{2\sqrt{a}}\right]$$
3. Use the Point-Slope Form for the Equation of the Tangent Line:

   The equation of the tangent line in point-slope form is:

   $$y-\sqrt{a}=m(x-a)$$
4. Substitute Values to Find the Equation:

Substitute the values of $(a),\left(\sqrt{a}\right),$ and (m) into the point-slope form to obtain the equation of the tangent line.

Example: Let's make a local linear approximation of the function $\left(y=\sqrt{x}\right)$ near the point $\left(a,\sqrt{a}\right)$ where $(a=4)$:

1. Calculate the derivative:

   $$\left(\frac{dy}{dx}=\frac{1}{2\sqrt{x}}\right)$$

2. Evaluate the derivative at $(x=4)$:

   $$\left(m=\frac{1}{2\sqrt{4}}=\frac{1}{4}\right)$$

3. Use the point-slope form:

   $$y-\sqrt{4}=\frac{1}{4}(x-4)$$

4. Simplify the equation:

   $$y-2=\frac{1}{4}(x-4)$$

This equation represents the tangent line to the curve $\left(y=\sqrt{x}\right)at\left(x=4\right)$.

- **Local Linear Approximation:** Now that we have the equation of the tangent line, we can use it to estimate the value of $\left(\sqrt{x}\right)$ near $(x=4)$. For example, if we want to approximate $\left(\sqrt{3}\right)$:

Substitute (x–3) into the above equation:

$$y-2=\frac{1}{4}(3-4)$$

$$y-2=-\frac{1}{4}$$

$$y=\frac{7}{4}$$

So, the local linear approximation of $\left(\sqrt{3}\right)$ near $(x=4)$ is approximately $\left(\frac{7}{4}\right)$.

Tangent lines and local linear approximations are valuable tools in calculus for estimating the behavior of functions near specific points, allowing us to make precise calculations and predictions in various scientific and engineering applications.

2.13. APPLICATIONS OF TANGENT LINES

The tangent function's versatile nature is a powerful tool for solving complex problems and making informed decisions across various disciplines, from the design of structures to the analysis of dynamic systems. This section delves into diverse real-world scenarios where the tangent function plays a pivotal role, illuminating its significance and impact on everyday lives.

2.13.1. Architecture and Engineering

Architects and engineers harness the mathematical prowess of the tangent function to design structures that strike a delicate balance between aesthetics and functionality. One of the most prevalent applications is determining the optimal inclination for ramps, roads, and staircases. By calculating the tangent of the desired inclination angle, designers ensure that these elements accommodate smooth movement while adhering to safety standards (Figure 2.7).

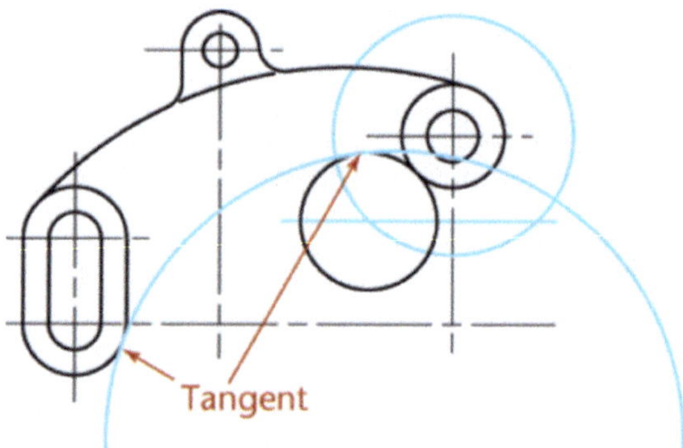

Figure 2.7. Tangent lines in architecture and engineering.

Source: Peachpit Press, Creative Commons License.

Consider a wheelchair ramp—a fundamental component of accessible design. The tangent function allows designers to precisely calculate the slope of the ramp, ensuring that individuals with mobility challenges can navigate it comfortably and safely. Similarly, road engineers employ the tangent function to design roads with appropriate inclines, minimizing the risk of accidents and enhancing driving conditions (Moroni et al., 2006).

2.13.2. Physics and Projectile Motion

The realm of physics witnesses the prowess of the tangent function in predicting the trajectories of projectiles and analyzing their motion through air. When an object is launched into the air, its path follows a parabolic curve influenced by launch angles and initial velocity.

By employing the tangent function, physicists can calculate the optimal launch angle for a projectile to achieve maximum range or height. The tangent of the launch angle provides insights into the relationship between vertical and horizontal components of motion, enabling accurate predictions of projectile behavior. This application finds relevance in fields ranging from ballistics and military strategy to sports and recreational activities.

2.13.3. Astronomy

In the cosmos, where celestial bodies dance across the expanse of space, the tangent function becomes a valuable tool for astronomers seeking to measure distances and angles. When observing objects in the night sky, astronomers often measure angles of elevation or depression to determine the positions of stars, planets, and other celestial objects (Figure 2.8).

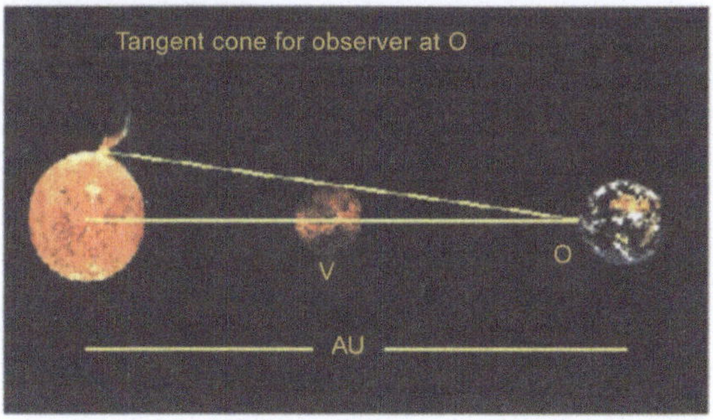

Figure 2.8. Tangent lines in astronomy.

Source: Sanath, Creative Commons License.

By employing the tangent function, astronomers calculate the distances to these celestial objects based on the measured angles and the Earth's radius. This technique, known as triangulation, allows astronomers to map out the positions of objects beyond the planet and gain insights into the vast cosmic landscape (Cohen, 1958).

2.13.4. Surveying and Navigation

Surveyors and navigators rely on the tangent function to map terrains, chart paths, and calculate distances. Surveyors use the tangent function to determine the vertical angles between points when conducting land surveys or creating maps. This information helps create accurate topographic maps that guide urban planning, construction, and infrastructure development.

In navigation, the tangent function calculates the line of sight between two points, which is crucial for determining distances and directions. Navigational tools like theodolite utilize the tangent function to measure angles and make precise calculations for navigation at sea, on land, or in the air.

2.13.5. Sound Engineering

Sound engineers and acoustic designers harness the power of the tangent function to manage sound waves in architectural and musical contexts. When designing concert halls or auditoriums, engineers use the tangent function to calculate the angles of reflection and dispersion of sound waves, ensuring optimal acoustics and sound quality for the audience.

Moreover, in musical instrument design, the tangent function influences the placement and design of soundboards and resonators. Engineers create instruments with enhanced tonal qualities and resonant frequencies by understanding the angles at which sound waves interact with surfaces (Chowning, 1971).

2.13.6. Photography and Art

Artists, photographers, and filmmakers employ the tangent function to capture and convey visual perspective in their creations. The concept of the tangent function finds expression in techniques such as foreshortening, where objects appear shorter in the viewer's perspective than in reality.

In visual arts, the tangent function helps artists accurately depict three-dimensional scenes on a two-dimensional canvas. By understanding the mathematical relationship between angles and proportions, artists can achieve realistic and visually compelling representations that engage the viewer's perception.

The applications of the tangent function extend far beyond the confines of mathematics textbooks, permeating a wide spectrum of real-world scenarios. Its ability to bridge angles and proportions, transform theoretical concepts into practical tools, and enable precise calculations makes it an indispensable asset in diverse fields.

From the design of accessible infrastructure to the prediction of celestial phenomena, from the optimization of trajectories to the creation of harmonious soundscapes, the tangent function stands as a testament to the profound connections between mathematics and the physical world. Its impact reverberates through architecture, physics, engineering, astronomy, navigation, art, and beyond, enriching lives with insights and solutions that merge abstract reasoning with tangible applications.

2.14. ADVANCED CONCEPTS IN TANGENT AND TANGENT LINES

This section delves into the intricate and multifaceted aspects of these fundamental mathematical tools, i.e., tangent functions. Beyond the basics, the following sections explore higher-order derivatives, curvature, radius of curvature, implicit differentiation, parametric equations, polar coordinates, and their applications in fields like physics, engineering, and optimization. Differential equations, numerical methods, and non-standard analysis further enhance our understanding of tangent-related phenomena. These advanced ideas deepen our mathematical toolbox, offering powerful insights into complex systems, dynamic processes, and optimization challenges across various disciplines.

2.14.1. Higher-Order Derivatives

2.14.1.1. Second and Higher Derivatives

In calculus, we start with the concept of the first derivative, which represents the rate of change of a function. Higher-order derivatives extend this idea by capturing how the rate of change itself changes. The second derivative denoted as $\left(f''(x)\right) or \left(\frac{d^2 f}{dx^2}\right)$, measures the rate at which the first derivative changes. Higher-order derivatives, such as the third $\left(f'''(x)\right)$ and fourth $\left(f''''(x)\right)$ derivatives, provide information about the change in the-change in the-change (and so on) of the function. In mathematical notation, higher-order derivatives are calculated by taking derivatives of derivatives.

2.14.1.2. Interpretation of Second Derivatives

The second derivative plays a crucial role in understanding the shape and behavior of a function. Here are key interpretations:

- **Concavity:** A function is concave up (or convex) if its second derivative is positive. This means that the function is curving upward. Conversely, if the second derivative is negative, the function is concave down, curving downward.
- **Inflection Points:** Inflection points are where the curvature of the function changes. At an inflection point, the second derivative changes sign from positive to negative or vice versa. These points often mark significant changes in the function's behavior.

Example: Consider the function $\left(f(x)=x^3-3x\right)$.

1. First Derivative: $\left(f'(x)=3x^2-3\right)$
2. Second Derivative: $\left(f''(x)=6x\right)$

i. The second derivative $\left(f''(x)\right)$ is positive when $(x>0)$, indicating that the function is concave up for positive (x).

ii. The second derivative $\left(f''(x)\right)$ is negative when $(x<0)$, indicating that the function is concave down for negative (x).

iii. The inflection point occurs at $(x=0)$ where the sign of the second derivative changes, marking the transition from concave down to concave up.

2.14.1.3. Concavity and Inflection Points

The concept of concavity and inflection points is essential in many applications, including physics and engineering. In physics, for example, understanding the concavity of a position-time graph can reveal information about an object's acceleration, while inflection points may indicate changes in motion. Higher-order derivatives go beyond the initial rate of change and provide deeper insights into a function's behavior. They are indispensable in various fields for analyzing curves, predicting trends, and optimizing processes.

2.14.2. Curvature and Radius of Curvature

Curvature is a geometric property of a curve that measures how much the curve deviates from being a straight line at a specific point. In calculus, the curvature is closely related to the second derivative of a function. The curvature (k) at a point on a curve is defined as the reciprocal of the radius of the osculating circle (the circle that best approximates the curve at that point) (Figure 2.9):

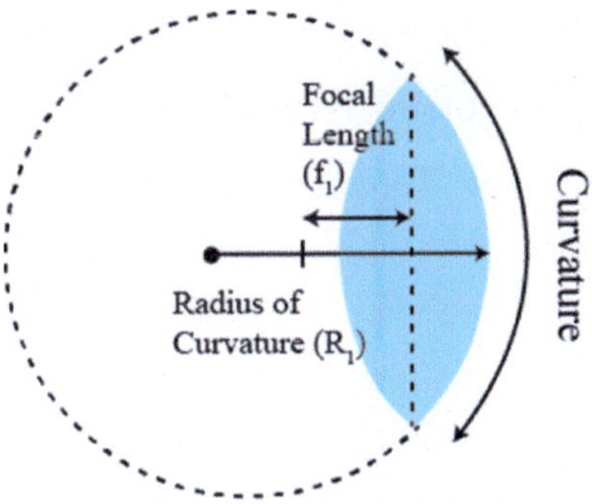

Figure 2.9. Curvature and radius of curvature.

Source: Interactive Mathematics Creative Commons License.

$$k = \frac{1}{R}$$

Where:

k is the curvature.

R is the radius of curvature.

Curvature is a measure of how sharply or gently a curve bends at a given point. Larger curvature values indicate sharper bends, while smaller values suggest gentler curves.

2.14.2.1. Calculating the Radius of Curvature

The radius of curvature R at a point on a curve is calculated using the formula:

$$R = \frac{1}{|k|}$$

Where:

R is the radius of curvature.

k is the curvature.

If you have the equation of the curve in parametric form $(x(t), y(t))$, the curvature can be computed using the following formula:

$$\text{k} = \frac{|x'y'' - y'x''|}{\left(x'^2 + y'^2\right)^{3/2}}$$

Where:

$x'(t)$ and $y'(t)$ are the first derivatives of (x) and (y) with respect to (t).

$x''(t)$ and $y''(t)$ are the second derivatives of (x) and (y) with respect to (t).

2.14.2.2. *Applications in Curve Analysis*

- **Road Design:** Engineers use curvature and radius of curvature to design roads and highways. Gentle curves have larger radii of curvature, while tight turns have smaller radii, affecting factors like speed limits and safety.
- **Car Racing:** In motorsports, understanding the curvature of a racetrack is crucial for drivers to optimize their racing lines and navigate turns efficiently.
- **Geometry:** Curvature is essential in differential geometry for studying curves and surfaces. It helps describe the local geometry of curves.
- **Optics:** In optics, curvature is used to describe the shape of lenses and mirrors, particularly in designing optical systems like telescopes and eyeglasses.

2.14.3. Implicit Differentiation

Implicit differentiation is a technique used in calculus to find the derivative of a function when the relationship between the variables is not expressed explicitly. In many cases, functions are defined implicitly as equations involving both dependent and independent variables. When it's difficult or impossible to solve explicitly for the dependent variable, implicit differentiation provides a way to find the derivative.

2.14.3.1. Derivatives of Implicit Functions

The process of implicit differentiation involves the following steps:

- **Differentiate Both Sides:** Start with the equation that defines the relationship between variables. Then, differentiate both sides of the equation with respect to the independent variable (usually (x)).
- **Apply Chain Rule:** When differentiating expressions involving the dependent variable, use the chain rule to differentiate it as if it were an explicit function of (x).
- **Isolate the Derivative:** After differentiation, isolate the derivative of the dependent variable on one side of the equation. This derivative represents $\left(\frac{dy}{dx}\right)$, where (y) is the dependent variable.
- **Solve for $\left(\frac{dy}{dx}\right)$:** If necessary, solve for $\left(\frac{dy}{dx}\right)$ to express it explicitly in terms of (x) and (y).

2.14.3.2. Applications in Geometry and Physics

Implicit differentiation finds applications in various fields, including geometry and physics:

- **Geometry:** Implicit differentiation can be applied to find the slope of a tangent line to a circle or an ellipse.
- **Physics:** In physics, many physical laws and relationships are defined implicitly. For instance, in thermodynamics, the ideal gas law $(PV = nRT)$ relates pressure (P), volume (V), temperature (T), and the number of moles (n), implicitly. By applying implicit differentiation, you can find the rate of change of one variable concerning another.

Example: Consider the equation of a circle:

$$x^2 + y^2 = r^2$$

Here, (x) and (y) are related implicitly by this equation, and you want to find $\left(\frac{dy}{dx}\right)$.

1. Differentiate both sides with respect to x:

   $$2x + 2yy' = 0$$

2. Apply the chain rule when differentiating y^2 (treat (y) as a function of (x):

   $$2x + 2y\frac{dy}{dx} = 0$$

3. Solve for $\left(\frac{dy}{dx}\right)$:

   $$\frac{dy}{dx} = -\frac{x}{y}$$

This is the derivative of \y\ with respect to \x\ for the circle equation. It represents the slope of the tangent line to the circle at any point (x, y).

Implicit differentiation is a powerful tool for handling equations where variables are intertwined, making it applicable in situations where explicit differentiation is not straightforward or possible. It is valuable in finding derivatives in various mathematical and scientific contexts.

2.14.4. Parametric Equations and Tangent Lines

2.14.4.1. Working with Parametric Equations

Parametric equations describe a curve using two or more related functions, typically in terms of a third variable (often denoted as $t)$. These equations provide a way to represent complex and dynamic shapes, such as curves in the plane or paths of moving objects. In parametric equations, each variable (x) and (y) is expressed as a function of (t), creating a set of equations that define the curve in terms of (t).

For example, a parametric equation for a point moving in a circular path of radius (r) is:

$$x(t) = r\cos(t)$$

$$y(t) = r\sin(t)$$

Here, (x) and (y) are defined as functions of (t), where (t) represents time or another parameter. These equations trace out a circle as (t) varies.

2.14.4.2. Tangent Lines to Parametric Curves

To find the equation of a tangent line to a parametric curve at a specific point, you can use the concept of derivatives with respect to (t). The steps involve:

- **Differentiating Parametric Equations:** Calculate the derivatives: $\left(\frac{dx}{dt}\right)$ and $\left(\frac{dy}{dt}\right)$ using the parametric equations.
- **The slope of the Tangent Line:** Find the slope of the tangent line at the point of interest by evaluating ($\frac{dy}{dt}$) divided by $\left(\frac{dx}{dt}\right)$ at that specific (t) value.
- **Equation of the Tangent Line:** Use the point-slope form of a line to write the equation of the tangent line at the point. The point on the tangent line is given by $(x(t_0), y(t_0))$, where (t_0) corresponds to the specific value of (t) for which you want to find the tangent.

2.14.4.3. Examples from Physics and Engineering

Parametric equations and tangent lines have applications in physics and engineering:

- **Physics:** Parametric equations are often used to describe the motion of objects in 2D and 3D space. For instance, in projectile motion, parametric equations are used to represent the path of a projectile. Tangent lines at specific points on the path help analyze velocity and acceleration.

- **Engineering:** Engineers use parametric equations to design curves and trajectories for objects such as aircraft, vehicles, and robots. Tangent lines are critical for understanding the orientation and dynamics of these systems at various points in their motion.

Example: Consider a particle following a parametric path given by:

$$x(t) = 2t^2$$

$$y(t) = 3t - 1$$

To find the equation of the tangent line at $(t=1)$, we follow these steps:

1. Calculate derivatives:

$$\left[\frac{dx}{dt} = 4t\right]$$

$$\left[\frac{dy}{dt} = 3\right]$$

2. Find the slope of the tangent line at $(t = 1)$ by evaluating the derivatives at that point:

$$m = \frac{dy}{dt}\Big|_{t=1} / \frac{dx}{dt}\Big|_{t=1} = \frac{3}{4}$$

3. The point on the curve at $(t = 1) is (2,2)$. Use the point-slope form to write the equation of the tangent line:

$$y - 2 = \frac{3}{4}(x - 2)$$

This equation represents the tangent line to the parametric curve at $(t=1)$.

Parametric equations are a versatile tool for describing dynamic curves, and tangent lines help analyze these curves at specific points.

2.14.5. Polar Coordinates and Tangents

Polar coordinates provide an alternative way to describe the position of a point in the plane using distance and angle rather than Cartesian coordinates (x, y). In polar coordinates, a point is defined by its distance from the origin (r) and the angle (θ) it makes with the positive x-axis.

Conversion from Cartesian to Polar:

$$r = \sqrt{x^2 + y^2}$$

$$\theta = \arctan\left(\frac{y}{x}\right)$$

Conversion from Polar to Cartesian:

$$x = r\cos(\theta)$$

$$y = r\sin(\theta)$$

2.14.5.1. Tangent Lines in Polar Coordinates

The following steps are used to find the equation of a tangent line to a curve represented in polar coordinates:

- **Differentiate (r) and $(\grave{e})$:** First, differentiate the polar equation with respect to (θ) to obtain $\left(\frac{dr}{d\theta}\right)$.
- **Derivative of (r) in Terms of $(\grave{e})$:** Express $\left(\frac{dr}{d\theta}\right)$ in terms of (r) and (θ) using the chain rule.
- **Slope of Tangent Line:** The derivative $\left(\frac{dr}{d\theta}\right)$ represents the rate of change of the radial distance (r) with respect to the angle (θ). The slope of the tangent line is given by:
- **Equation of Tangent Line:** Using the point-slope form of a line, write the equation of the tangent line at the desired point.

2.14.5.2. Applications in Engineering and Astronomy

- **Engineering:** Polar coordinates are used in engineering for analyzing and designing systems with rotational symmetry. For example, in designing gears, turbines, and antennas, engineers use polar coordinates to describe the geometry and orientation of components.
- **Astronomy:** Polar coordinates are crucial in astronomy for locating celestial objects in the night sky. Astronomers use polar coordinates to describe the position of stars and other celestial bodies, including their distance from Earth and their angular positions.

Example: Consider the polar equation $\left(r = 2 + 3\cos(\theta)\right)$ to find the equation of the tangent line at a specific point, say $\left(\theta = \frac{\pi}{6}\right)$:.

Solution:

1. Differentiate (r) with respect to (θ):

$$\frac{dr}{d\theta} = -3\sin(\theta)$$

2. Express the slope of the tangent line:

$$\left[m = \frac{-3\sin\left(\frac{\pi}{6}\right)\sin(\theta) + \left(2 + 3\cos(\theta)\right)\cos(\theta)}{-3\sin\left(\frac{\pi}{6}\right)\cos(\theta) - \left(2 + 3\cos(\theta)\right)\sin(\theta)} \right]$$

3. Simplify the expression for (m) and substitute $\left(\theta = \frac{\pi}{6}\right)$ to find the slope.
4. Use the point-slope form to write the equation of the tangent line at $\left(\theta = \frac{\pi}{6}\right)$.

In engineering and astronomy, understanding polar coordinates and tangent lines is essential for solving problems involving rotational symmetry and locating objects in a polar coordinate system.

2.14.6. Advanced Applications

2.14.6.1. Differential Equations and Tangent Fields

Differential equations involving tangent fields are a powerful tool for modeling dynamic systems and understanding how they change over time. Tangent fields are essentially collections of tangent vectors at each point in a space. In this context:

- **Ordinary Differential Equations (ODEs):** Tangent lines represent the direction in which a solution to an ODE is evolving at a given point. Solving ODEs with tangent fields helps predict the behavior of dynamic systems.
- Partial Differential Equations (PDEs)**:** In PDEs, tangent fields become tangent planes or higher-dimensional surfaces. They are used to study phenomena that vary in multiple dimensions, such as heat diffusion, wave propagation, or fluid flow.

2.14.6.2. Numerical Methods for Tangent Line Approximations

Numerical methods, like Euler's method and the Runge-Kutta method, use tangent line approximations to estimate solutions to differential equations. These methods work by iteratively applying tangent lines to approximate the next point on a curve, allowing for step-by-step numerical solutions to ODEs and PDEs. They are essential in fields such as physics, engineering, and computer science for simulating complex systems.

2.14.6.3. Tangent Lines in Optimization Problems

Tangent lines are employed in optimization problems to find the local extrema (maxima or minima) of functions. The idea is to start at an initial point and follow the direction of the tangent line to reach an optimal solution iteratively. This technique is used in mathematical optimization, engineering design, and machine learning. For instance, in calculus-based optimization:

- **Gradient Descent:** Tangent lines help update the parameters of a model iteratively to minimize a cost function. The negative gradient (tangent line) points in the direction of the steepest decrease in the cost.
- **Newton's Method:** Tangent lines are used to approximate the roots of a function by iteratively finding the x-intercept of the

tangent line. This method is particularly efficient for functions with well-behaved local extrema.

Example: Consider a classic optimization problem: finding the minimum of the function $(f(x) = x^2 - 4x + 5)$. We can use tangent lines iteratively to approximate the minimum:

1. Start at an initial guess, say $(x_0 = 0)$.
2. Calculate the derivative $(f'(x))$ and evaluate it at (x_0).
3. The tangent line's slope at (x_0) is $(f'(x_0))$.
4. Find the x-intercept of the tangent line as $\left(x_1 = x_0 - \frac{f(x_0)}{f'(x_0)}\right)$.
5. Repeat steps 2–4 until convergence (i.e., until $(f'(x_n))$ is close to zero).

The sequence of (x_n) values approach the minimum of $(f(x))$, which is $(x=2)$ in this case.

Tangent lines have advanced applications in differential equations, numerical methods, and optimization. They are instrumental in modeling dynamic systems, numerically solving differential equations, and finding optimal solutions in various fields of science and engineering.

2.14.7. Non-Standard Analysis

Non-standard analysis is a mathematical framework that extends classical real analysis by introducing infinitesimals and hyperreal numbers. It was pioneered by Abraham Robinson in the mid-20th century as a way to rigorously deal with concepts of limits, continuity, and derivatives in a manner that includes infinitesimal quantities. Non-standard analysis provides an alternative foundation for calculus and opens up new perspectives on mathematical analysis.

2.14.7.1. Infinitesimals and Hyperreal Numbers

In non-standard analysis, infinitesimals are quantities that are infinitely small but not equal to zero. They are denoted by symbols like (ϵ) or $(\ddot{\text{a}})$ and are used to represent infinitesimal changes in variables. For example, an infinitesimal increment in (x) is represented as $(\ddot{\text{A}}x = \epsilon)$.

Hyperreal numbers are an extension of the real numbers that include infinitesimals and infinite numbers. They are represented as $(a+\epsilon)$, where (a) is a real number and (ϵ) is an infinitesimal. Hyperreal numbers maintain the usual algebraic properties of real numbers but allow for operations involving infinitesimal and infinite quantities.

2.14.7.2. Tangent Lines in Non-Standard Calculus

In non-standard calculus, tangent lines are approached in a manner that incorporates infinitesimals. The key concept is the notion of a hyperreal number representing the slope of a tangent line at a point on a curve. This is achieved through a process called the "transfer principle," which allows statements in the hyperreal number system to be translated back into the real number system.

- **Hyperreal Tangent Lines:** In the context of non-standard analysis, a tangent line to a curve at a point is defined using a hyperreal number as the slope. This hyperreal slope takes into account infinitesimal changes in the function near the point of tangency.
- **Transfer Principle:** The transfer principle is a fundamental idea in non-standard analysis. It asserts that if a statement is true for the hyperreal numbers, it corresponds to a true statement about the real numbers. This principle bridges the gap between the real number system and the hyperreal number system, enabling the study of tangent lines with infinitesimal precision.

The non-standard analysis provides a unique and powerful perspective on calculus, allowing mathematicians to work with infinitesimals and study tangent lines in a way that aligns with the intuitive notions of calculus without sacrificing rigor. This framework has applications in various areas of mathematics, including mathematical physics and functional analysis.

Activity 1.1

Imagine you are an architect tasked with designing a building. The building has a sloped roof, and you need to calculate the angle of inclination for proper drainage. How would you use the tangent function and its properties to determine this angle effectively?

2.15. SUMMARY

This chapter, "The Tangent," is a comprehensive exploration of the tangent function in trigonometry. It starts by defining the tangent function (tan θ) and its role in right triangles. The chapter covers the fundamental properties of a tangent, including its periodicity, asymptotes, range, and graphical behavior. It provides insight into graphing the tangent function and solving equations involving it. Additionally, the chapter delves into advanced concepts like higher-order derivatives, curvature, implicit differentiation, and parametric equations. It's a crucial foundation for understanding the tangent function's significance and versatility in mathematics and real-world scenarios.

REFERENCES

1. Adler, R., (1973). A study of locking phenomena in oscillators. *Proceedings of the IEEE, 61*(10), 1380–1385.
2. Asgarli, S., & Ghioca, D., (2023). Tangent-filling plane curves over finite fields. *Bulletin of the Australian Mathematical Society, 10*, 1–15.
3. Ba, J., Hu, P., Tan, W., Müller, T. M., & Fu, L. Y., (2021). Brittle mineral prediction based on rock-physics modelling for tight oil reservoir rocks. *Journal of Geophysics and Engineering, 18*(6), 970–983.
4. Bivens, I. C., (1986). What a tangent line is when it isn't a limit. *The College Mathematics Journal, 17*(2), 133–143.
5. Brown, R. S., (1921). Use of the tangent chart for solving transmission line problems. *Journal of the American Institute of Electrical Engineers, 40*(11), 854–864.
6. Cafiso, S., Di Graziano, A., Di Silvestro, G., La Cava, G., & Persaud, B., (2010). Development of comprehensive accident models for two-lane rural highways using exposure, geometry, consistency, and context variables. *Accident Analysis & Prevention, 42*(4), 1072–1079.
7. Cavanagh, M., (2008). Trigonometry from a different angle. *Australian Mathematics Teacher, the, 64*(1), 25–30.
8. Chang, L., & Wei, W., (2022). Numerical study on the effect of tangential intake on vortex dropshaft assessment using pressure distributions. *Engineering Applications of Computational Fluid Mechanics, 16*(1), 1100–1110.
9. Cheung, Y. T., Foo, Y. L., Shwe, M., Tan, Y. P., Fan, G., Yong, W. S., & Chan, A., (2014). Minimal clinically important difference (MCID) for the functional assessment of cancer therapy: Cognitive function (FACT-Cog) in breast cancer patients. *Journal of Clinical Epidemiology, 67*(7), 811–820.
10. Chowning, J. M., (1971). The simulation of moving sound sources. *Journal of the Audio Engineering Society, 19*(1), 2–6.
11. Cohen, M. H., (1958). Radio astronomy polarization measurements. *Proceedings of the IRE, 46*(1), 172–183.
12. Doghri, I., & Ouaar, A., (2003). Homogenization of two-phase elasto-plastic composite materials and structures: Study of tangent operators, cyclic plasticity, and numerical algorithms. *International Journal of Solids and Structures, 40*(7), 1681–1712.

13. Du, S. S., Hou, K., Salakhutdinov, R. R., Poczos, B., Wang, R., & Xu, K., (2019). Graph neural tangent kernel: Fusing graph neural networks with graph kernels. *Advances in Neural Information Processing Systems, 32*, 189–196.

14. Engheta, N., (2002). An idea for thin subwavelength cavity resonators using metamaterials with negative permittivity and permeability. *IEEE Antennas and Wireless Propagation Letters, 1*, 10–13.

15. Ernest, P., (1997). Mathematical activity and rhetoric: A semiotic analysis of an episode of mathematical activity. *Philosophy of Mathematics Education Journal, 10*, 523–525.

16. Extrand, C. W., (2016). Uncertainty in contact angle measurements from the tangent method. *Journal of Adhesion Science and Technology, 30*(15), 1597–1601.

17. Fahami, S. M. H., Zamzuri, H., Mazlan, S. A., & Saruchi, S. A., (2014). The variable steering ratio for vehicle steer by wire system using hyperbolic tangent method. *Applied Mechanics and Materials, 575*, 781–784.

18. Fellin, W., & Ostermann, A., (2003). Using constitutive models of the rate type in implicit finite-element calculations: Error-controlled stress update and consistent tangent operator. In: *Advanced Mathematical and Computational Geomechanics* (Vol. 2, pp. 211–237). Berlin, Heidelberg: Springer Berlin Heidelberg.

19. Fenner, B. J., Ting, D. S., Tan, A. C., Teo, K., Chan, C. M., Mathur, R., & Cheung, C. M. G., (2020). Real-world treatment outcomes of age-related macular degeneration and polypoidal choroidal vasculopathy in Asians. *Ophthalmology Retina, 4*(4), 403–414.

20. Fujiwara, M., Wright, J. S., Manney, G. L., Gray, L. J., Anstey, J., Birner, T., & Zou, C. Z., (2017). Introduction to the SPARC reanalysis intercomparison project (S-RIP) and overview of the reanalysis systems. *Atmospheric Chemistry and Physics, 17*(2), 1417–1452.

21. Göküzüm, F. S., & Keip, M. A., (2018). An algorithmically consistent macroscopic tangent operator for FFT-based computational homogenization. *International Journal for Numerical Methods in Engineering, 113*(4), 581–600.

22. Greco, L., & Cuomo, M., (2015). Consistent tangent operator for an exact Kirchhoff rod model. *Continuum Mechanics and Thermodynamics, 27*, 861–877.

23. Gunasekaran, N., & Joo, Y. H., (2020). Nie–tan fuzzy method of fault-tolerant wind energy conversion systems via sampled-data control. *IET Control Theory & Applications, 14*(11), 1516–1523.
24. Guo, J., Ge, Y., Hao, L., Tan, J., Peng, Z., & Zhang, C., (2015). Comparison of real-world fuel economy and emissions from parallel hybrid and conventional diesel buses fitted with selective catalytic reduction systems. *Applied Energy, 159*, 433–441.
25. Hilfiker, J. N., Singh, N., Tiwald, T., Convey, D., Smith, S. M., Baker, J. H., & Tompkins, H. G., (2008). Survey of methods to characterize thin absorbing films with spectroscopic ellipsometry. *Thin Solid Films, 516*(22), 7979–7989.
26. Homma, M., (1994). Plane curves whose tangent lines at collinear points are concurrent. *Geometriae Dedicata, 53*, 287–296.
27. Horemuz, M., (1999). Error calculation in maritime delimitation between states with opposite or adjacent coasts. *Marine Geodesy, 22*(1), 1–17.
28. Jehel, P., Léger, P., & Ibrahimbegovic, A., (2014). Initial versus tangent stiffness-based Rayleigh damping in inelastic time history seismic analyses. *Earthquake Engineering & Structural Dynamics, 43*(3), 467–484.
29. Kuwata, M., (2005). Twenty-eight double tangent lines of a plane quartic curve with an involution and the Mordell-Weil lattices. *Rikkyo Daigaku Sugaku Zasshi, 54*(1), 17–32.
30. Levy, S. G., & Doherty, M. F., (1986). A simple exact method for calculating tangent pinch points in multicomponent nonideal mixtures by bifurcation theory. *Chemical Engineering Science, 41*(12), 3155–3160.
31. Lu, X., & Hsu, C. T., (2007). Tangent Poisson's ratio of high-strength concrete in triaxial compression. *Magazine of Concrete Research, 59*(1), 69–77.
32. Maknun, C. L., Rosjanuardi, R., & Jupri, A., (2019). From ratios of right triangle to unit circle: An introduction to trigonometric functions. In: *Journal of Physics: Conference Series* (Vol. 1157, pp. 221–224). IOP Publishing.
33. Marmur, A., (2006). Soft contact: Measurement and interpretation of contact angles. *Soft Matter, 2*(1), 12–17.
34. Mednykh, A. D., (2002). Trigonometric identities and geometrical

inequalities for links and knots. In: *Proceedings of the Third Asian Mathematical Conference 2000* (Vol. 2, pp. 352–368).

35. Melrose, R., Barreto, A. S., & Vasy, A., (2014). Analytic continuation and semiclassical resolvent estimates on asymptotically hyperbolic spaces. *Communications in Partial Differential Equations, 39*(3), 452–511.
36. Mokhtarian, F., & Mackworth, A. K., (1992). A theory of multiscale, curvature-based shape representation for planar curves. *IEEE Transactions on Pattern Analysis and Machine Intelligence, 14*(8), 789–805.
37. Mondal, K., & Pramanik, S., (2015). Neutrosophic tangent similarity measure and its application to multiple attribute decision making. *Neutrosophic Sets and Systems, 9*, 80–87.
38. Moroni, L., De Wijn, J. R., & Van, B. C. A., (2006). 3D fiber-deposited scaffolds for tissue engineering: Influence of pores geometry and architecture on dynamic mechanical properties. *Biomaterials, 27*(7), 974–985.
39. Muthusamy, S., Wang, S., & Chung, D. D. L., (2010). Unprecedented vibration damping with high values of loss modulus and loss tangent, exhibited by cement–matrix graphite network composite. *Carbon, 48*(5), 1457–1464.
40. Na, H. M., Yang, H. K., & Hwang, J. M., (2021). Visual field ratio according to distance in the near tangent screen test. *Journal of the Korean Ophthalmological Society, 62*(2), 261–265.
41. Owen, E., & Sweller, J., (1985). What do students learn while solving mathematics problems? *Journal of Educational Psychology, 77*(3), 272.
42. Papneja, N., Zaidi, A., Chalchal, H., Moser, M., Tan, K., Olson, C., & Ahmed, S., (2019). Comparisons of outcomes of real-world patients with advanced pancreatic cancer treated with FOLFIRINOX versus gemcitabine and nab-paclitaxel: A population-based cohort study. *Pancreas, 48*(7), 920–926.
43. Parvizi, J., Tan, T. L., Goswami, K., Higuera, C., Della, V. C., Chen, A. F., & Shohat, N., (2018). The 2018 definition of periprosthetic hip and knee infection: An evidence-based and validated criteria. *The Journal of Arthroplasty, 33*(5), 1309–1314.
44. Poon, K. K., Soon, S., Wong, M. E., Kaur, S., Khaw, J., Ng, Z., &

Tan, C. S., (2014). What is school like? Perspectives of Singaporean youth with high-functioning autism spectrum disorders. *International Journal of Inclusive Education, 18*(10), 1069–1081.

45. Rifai, S., Dauphin, Y. N., Vincent, P., Bengio, Y., & Muller, X., (2011). The manifold tangent classifier. *Advances in Neural Information Processing Systems, 24*, 789–790.
46. Rius, J., (1993). Derivation of a new tangent formula from Patterson-function arguments. *Acta Crystallographica Section A: Foundations of Crystallography, 49*(3), 406–409.
47. Rivero, A., (1997). *Introduction to the Tangent Grupoid* (Vol. 762, No. 45, pp. 97522–97525). arXiv preprint dg-ga/9710026.
48. Roberts, A. W., (2001). Design considerations and performance evaluation of screw conveyors. *Proceedings of the Beltcon, 11*(11), 263–270.
49. Rohlf, F. J., (1999). Shape statistics: Procrustes superimpositions and tangent spaces. *Journal of Classification, 16*, 197–223.
50. Rosjanuardi, R., & Jupri, A., (2020). Didactical design on drawing and analyzing trigonometric functions graph through a unit circle approach. *International Electronic Journal of Mathematics Education, 15*(3), 468–470.
51. Roxburgh, I. W., & Vorontsov, S. V., (2003). The ratio of small to large separations of acoustic oscillations as a diagnostic of the interior of solar-like stars. *Astronomy & Astrophysics, 411*(2), 215–220.
52. Rubino, E., & Ioppolo, T., (2016). Young's modulus and loss tangent measurement of polydimethylsiloxane using an optical lever. *Journal of Polymer Science Part B: Polymer Physics, 54*(7), 747–751.
53. Sasaki, S., (1958). On the differential geometry of tangent bundles of Riemannian manifolds. *Tohoku Mathematical Journal, Second Series, 10*(3), 338–354.
54. Shimizu, Y., & Cundall, P. A., (2001). Three-dimensional DEM simulations of bulk handling by screw conveyors. *Journal of Engineering Mechanics, 127*(9), 864–872.
55. Skaar, D. L., (1999). Predicting induction motor performance using a unit circle. In: *Proceedings: Electrical Insulation Conference and Electrical Manufacturing and Coil Winding Conference (Cat. No. 99CH37035)* (Vol. 2, pp. 421–426). IEEE.
56. Starman, B., Halilovič, M., Vrh, M., & Štok, B., (2014). Consistent

tangent operator for cutting-plane algorithm of elasto-plasticity. *Computer Methods in Applied Mechanics and Engineering, 272*, 214–232.

57. Starr, R. R., (1973). Using trigonometry in earth science astronomy labs. *Journal of Geological Education, 21*(4), 148–151.
58. Steinmann, P., Hossain, M., & Possart, G., (2012). Hyperelastic models for rubber-like materials: Consistent tangent operators and suitability for Treloar's data. *Archive of Applied Mechanics, 82*, 1183–1217.
59. Sweller, J., (1988). Cognitive load during problem solving: Effects on learning. *Cognitive Science, 12*(2), 257–285.
60. Szabó, L., & Jonas, J. J., (1995). Consistent tangent operator for plasticity models based on the plastic strain rate potential. *Computer Methods in Applied Mechanics and Engineering, 128*(3, 4), 315–323.
61. Talkokul, E. N., (2017). Research on the trigonometry as a main function of sine, secant, tangent and formula of tan and cot are inverse. *International Journal of Mathematics Trends and Technology, 50*, 147–152.
62. Tan, J. W. L., Valerio, C., Barnes, E. H., Turner, P. J., Van, A. P. A., Kakakios, A. M., & Group, B. E. A. T. B. S. (2017). A randomized trial of egg introduction from 4 months of age in infants at risk for egg allergy. *Journal of Allergy and Clinical Immunology, 139*(5), 1621–1628.
63. Tan, T., Mills, G., Papadonikolaki, E., & Liu, Z., (2021). Combining multi-criteria decision making (MCDM) methods with building information modelling (BIM): A review. *Automation in Construction, 121*, 103451.
64. Thoniyot, P., Tan, M. J., Karim, A. A., Young, D. J., & Loh, X. J., (2015). Nanoparticle–hydrogel composites: Concept, design, and applications of these promising, multi-functional materials. *Advanced Science, 2*(1, 2), 1400010.
65. Troyanovich, S. J., Harrison, D. E., Harrison, D. D., Holland, B., & Janik, T. J., (1998). Further analysis of the reliability of the posterior tangent lateral lumbar radiographic mensuration procedure: Concurrent validity of computer-aided x-ray digitization. *Journal of Manipulative and Physiological Therapeutics, 21*(7), 460–467.
66. Van, V. C. N., Lam, V. T., Dung, N. T., Yen, L. M., Minh, N. N. Q., Hung, L. M., & Van, T. L., (2020). The natural history and transmission

potential of asymptomatic severe acute respiratory syndrome coronavirus 2 infection. *Clinical Infectious Diseases, 71*(10), 2679–2687.

67. Vidal, C. A., Lee, H. S., & Haber, R. B., (1991). The consistent tangent operator for design sensitivity analysis of history-dependent response. *Computing Systems in Engineering, 2*(5, 6), 509–523.
68. Virbhadra, K. S., (2009). Relativistic images of Schwarzschild black hole lensing. *Physical Review D, 79*(8), 083004.
69. Watson, J., Beswick, K., & Brown, N., (2012). Estimating tree heights for middle school. In: *Educational Research and Professional Learning in Changing Times* (Vol. 2, pp. 201–210). Brill.
70. Watson, J., Brown, N., Wright, S., & Skalicky, J., (2011). Discovery a middle-school classroom inquiry: Estimating the height of a tree. *Australian Mathematics Teacher, 67*(2), 14–21.
71. Wu, J., Chen, L., Wang, Y., Tan, W., & Huang, Z., (2019). Prognostic value of aspartate transaminase to alanine transaminase (De Ritis) ratio in solid tumors: A pooled analysis of 9,400 patients. *OncoTargets and Therapy, 12*, 5201.
72. Xiao, F., Honma, Y., & Kono, T., (2005). A simple algebraic interface capturing scheme using hyperbolic tangent function. *International Journal for Numerical Methods in Fluids, 48*(9), 1023–1040.
73. Yee, J., Carson, J. M., Hajarizadeh, B., Hanson, J., O'Beirne, J., Iser, D., & Miczkova, S., (2022). High effectiveness of broad access direct-acting antiviral therapy for hepatitis C in an Australian real-world cohort: The REACH-C study. *Hepatology Communications, 6*(3), 496–512.
74. Yusha'u, M. A., (2013). Difficult topics in junior secondary school mathematics: Practical aspect of teaching and learning trigonometry. *Scientific Journal of Pure and Applied Sciences, 2*(4), 161–174.
75. Zamanlooy, B., & Mirhassani, M., (2013). Efficient VLSI implementation of neural networks with hyperbolic tangent activation function. *IEEE Transactions on Very Large-Scale Integration (VLSI) Systems, 22*(1), 39–48.
76. Zayed, E. M. E., & Gepreel, K. A., (2009). The (G′/G)-expansion method for finding traveling wave solutions of nonlinear partial differential equations in mathematical physics. *Journal of Mathematical Physics, 50*(1), 587–590.

77. Zhang, H., Li, N., Wu, J., Su, L., Chen, X., Lin, B., & Luo, H., (2013). Galangin inhibits proliferation of HepG2 cells by activating AMPK via increasing the AMP/TAN ratio in a LKB1-independent manner. *European Journal of Pharmacology, 718*(1–3), 235–244.

CHAPTER 3

Sine and Cosine

Contents

3.1. INTRODUCTION

Sine $(\sin)$ and cosine $(\cos)$ functions are fundamental trigonometric functions that describe the relationships between angles and the sides of right triangles. They are defined using the unit circle, where the radius is one, and the angle is measured from the positive x-axis in a counterclockwise direction. These functions have important mathematical properties and find extensive applications in various fields, including physics, engineering, and signal processing.

3.1.1. Sine Function $(\sin)$

The sine of an angle (è) (in radians) is defined as the y-coordinate of the point where the terminal side of the angle intersects the unit circle (Figure 3.1).

- **Mathematical Representation:** $\sin(\theta) = \frac{y}{r}$
- **Range:** The sine function's values range from -1 to 1: $-1 \leq \sin(\theta) \leq 1$
- **Graphical Representation:** The sine function's graph is a periodic wave that oscillates between -1 and 1, with a period of (2π) radians (360 degrees).

Examples:

1. $\sin(0) = 0$; because the angle is on the positive x-axis.
2. $\sin\left(\frac{\pi}{2}\right) = 1$; because the angle is at 90 degrees.
3. $\sin(\pi) = 0$; because the angle is on the negative x-axis.

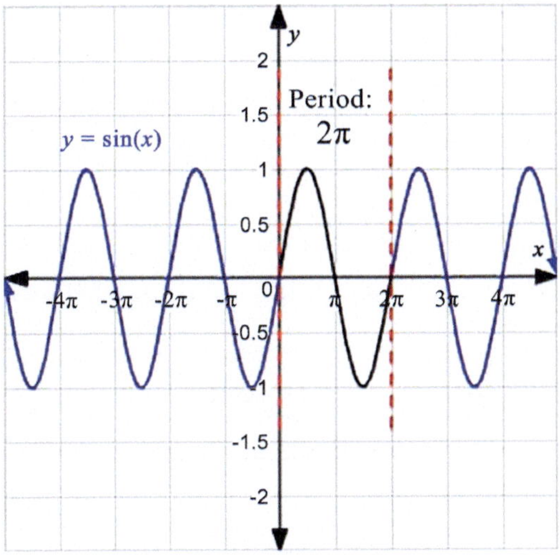

Figure 3.1. Graphical representation of sine function.

Source: Varsity tutors, Creative Commons License.

3.1.2. Cosine Function (cos)

The cosine of an angle (θ) (in radians) is defined as the x-coordinate of the point where the terminal side of the angle intersects the unit circle.

- **Mathematical Representation:** $\cos(\theta) = \frac{x}{r}$
- **Range:** The cosine function's values also range from -1 to 1: $-1 \leq \cos(\theta) \leq 1$
- **Graphical Representation:** The cosine function's graph is similar to the sine function but shifted horizontally by $\left(\frac{\pi}{2}\right)$ radians (90 degrees). It also has a period of (2π) radians.

Examples:

1. $\cos(0) = 1$; because the angle is on the positive x-axis.

2. $\cos\left(\frac{\pi}{2}\right) = 0$; because the angle is at 90 degrees.

3. $\cos(\pi) = -1$; because the angle is on the negative x-axis.

3.2. SINE AND COSINE IN A TRIANGLE

Figure 3.2 shows the construction of a perpendicular line from point A to line OB. Now consider the ratios of the lengths of lines AB and OB to the length of the hypotenuse OA in triangle OAB in the figure.

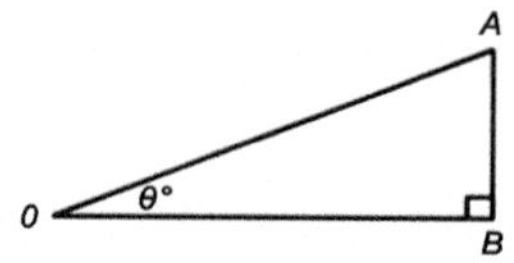

Figure 3.2. Illustration of an OAB triangle.

Source: P. Abott, Creative Commons License.

For a given fixed angle θ°, the ratio remains constant (equal to tan θ°) regardless of the position of point A. The ratio of opposite to hypotenuse is commonly referred to as the sine of the angle θ° and is denoted as sin θ° (Leindler, 2002) (Figure 3.3).

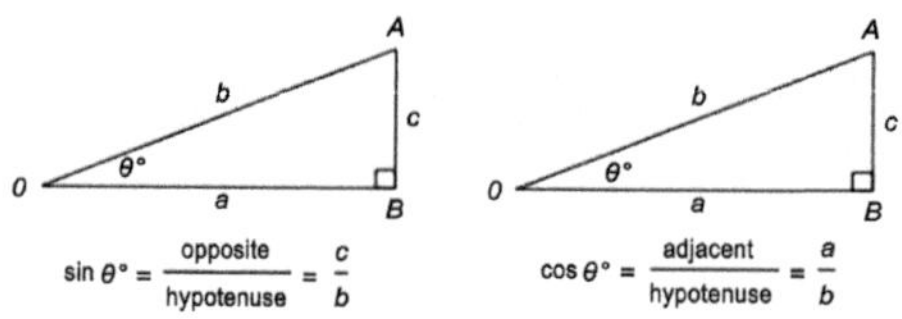

Figure 3.3. Sine and Cosine for OAB triangle.

Source: P. Abott, Creative Commons License.

Likewise, the ratio adjacent to the hypotenuse remains constant for the angle θ°. This ratio is called the cosine of the angle θ° and is written as cos θ° (Takagi et al., 1991).

Thus,

$$sin\theta° = \frac{opposite}{hypotenuse} = \frac{c}{b}, cos\theta° = \frac{adjacent}{hypotenuse} = \frac{a}{b}.$$

3.3. SINE AND COSINE FUNCTIONS IN A CIRCLE

In trigonometry, sine (sin) and cosine (cos) are used to find a geometric interpretation when applied to a unit circle (a circle with a radius of 1). This visual approach provides an intuitive understanding of these functions and extends their usefulness to angles beyond the scope of right triangle.

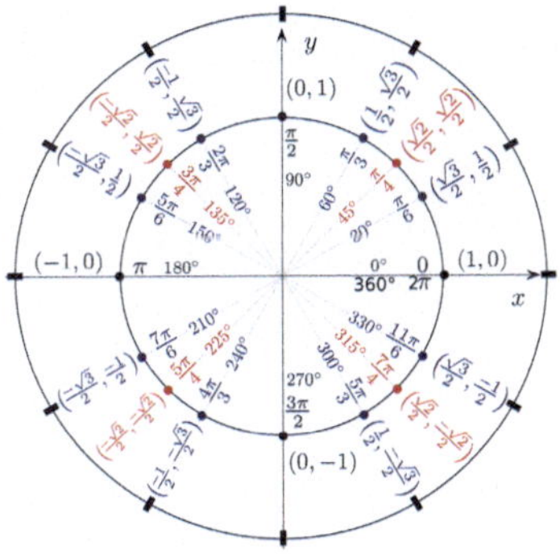

Figure 3.4. Sine and Cosine in a circle.

Source: Diane Rocha, Creative Commons License.

As we measure an angle (θ) counterclockwise from the positive x-axis in this unit circle, we reach a point on the circle's circumference. The x-coordinate of this point is defined as $\cos(\theta)$, and the y-coordinate is defined as $\sin(\theta)$. These coordinates give us a unique insight into the relationships between angles and the unit circle.

- Example:

When $(\theta)$is$\left(\frac{\pi}{4}\right)$ radians (45 degrees), the point on the unit circle has coordinates $\left(\cos\left(\frac{\pi}{4}\right), \sin\left(\frac{\pi}{4}\right)\right)$. Calculating this, we find

$$\left(\cos\left(\frac{\pi}{4}\right) = \frac{1}{\sqrt{2}}\right) \text{and} \left(\sin\left(\frac{\pi}{4}\right) = \frac{1}{\sqrt{2}}\right).$$

This visual representation of sine and cosine on a unit circle enhances our geometric understanding of these functions and serves as a foundation for trigonometric concepts. These concepts are widely applied in fields like physics, engineering, and geometry, where angles and circles play a crucial role in problem-solving and analysis.

The following paragraph provides a concise introduction to the sine and cosine ratios. A circular shape with a radius of 1 unit is constructed with its center at the origin point (O). Then, the radius line segment (OP) is constructed at an angle θ° to the x-axis in a counterclockwise direction, as depicted in Figure 3.5 (Alotaibi et al., 2020).

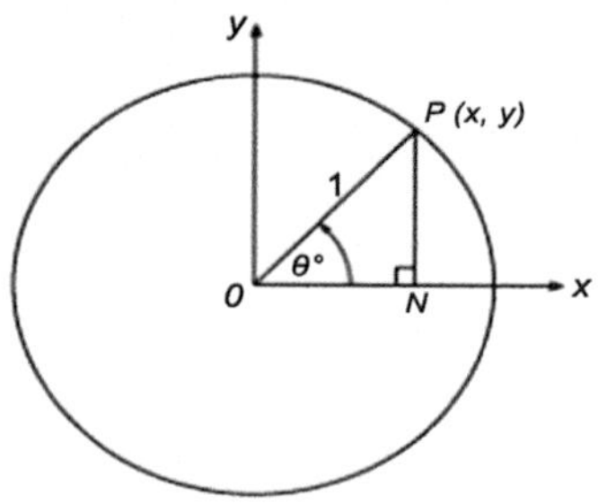

Figure 3.5. Circle with origin O at the center.

Source: P. Abott, Creative Commons License.

Next, consider the point P with coordinates (x, y). The definitions of sine and cosine are given by the equations sin θ° = y and cos θ° = x, respectively. An arrow in Figure 3.4 indicates the angle θ° and highlights the convention of measuring angles counterclockwise. Additionally, it is important to consider the following characteristics of these trigonometric functions:

i. In the triangle OPN, angle OPN = (90 – θ)°, and

$$sin\left(90-\theta\right)° = \frac{opposite}{hypotenuse} = \frac{x}{1} = cos\theta°$$

That is $sin\left(90-\theta\right)° = cos\theta°$

Similarly, $cos\left(90-\theta\right)° = sin\theta°$

ii. Using Pythagoras theorem on triangle OPN gives $x^2 + y^2 = 1$. Therefore, $\sin^2\theta + \cos^2\theta = 1$,

The equation $\sin^2\theta + \cos^2\theta = 1$, is referred to as the Pythagorean identity. Finding the values of the sine and cosine of angles is similar to finding the tangent of an angle. You can use the inverse sine and cosine functions, denoted as $\sin^{-1}$ and $\cos^{-1}$, respectively, to determine the inverse values of sine and cosine. Similarly, the inverse tangent function, $\tan^{-1}$, is used to ascertain the inverse tangent (Bass et al., 1975).

3.4. USING THE SINE AND COSINE

Two fundamental functions of trigonometry, sine, and cosine, are widely used to analyze angles, distances, and cyclic phenomena in various fields. These functions are based on the principles of right triangles and the unit circle. A thorough exploration of the basics of trigonometry reveals how sine and cosine seamlessly integrate into real-world situations. (Mahmud, 1990).

3.4.1. Angle Measurement and Geometry

Sine and cosine are indispensable tools for measuring angles and distances in geometry. Given an angle θ and the lengths of sides in a right triangle, sine and cosine can determine the lengths of the opposite and adjacent sides, respectively. This knowledge finds practical use in land surveying, architecture, and construction, where accurate measurements are vital for designing buildings, roads, and other structures (Xie et al., 2018).

Let's consider a triangle in a two-dimensional space with vertices A, B, and C. The coordinates of these vertices are as follows: A(0, 0), B(a, b), and C(c, d), where a, b, c, and d are real numbers.

The angle θ formed by the line segments AB and AC can be expressed using the dot product and magnitudes of vectors AB and AC as follows:

$$\cos(\theta) = \frac{\overrightarrow{AB} \cdot \overrightarrow{AC}}{|\overrightarrow{AB}| \cdot |\overrightarrow{AC}|}$$

Where:

$\cos(\theta)$ Represents the cosine of angle θ.

$\overrightarrow{AB}$ Represents the vector from A to B.

$\overrightarrow{AC}$ Represents the vector from A to C.

$(\cdot)$ Denotes the dot product of vectors.

$|\overrightarrow{AB}|$ Signifies the magnitude (length) of vector AB.

$\left|\overrightarrow{AC}\right|$ Denotes the magnitude (length) of vector AC.

It's a fundamental formula used in vector geometry and can be applied to various fields, including physics and engineering problems involving vector analysis.

3.4.2. Navigation and Astronomy

Sine and cosine applications extend beyond Earth's surface, aiding navigators and astronomers alike. In navigation, these functions help determine distances and angles between landmarks, enabling precise calculations of global positions. Astronomers rely on sine and cosine to model celestial motions, predict eclipses, and chart the paths of planets and stars across the sky (Peirce, 1845).

Consider celestial navigation, where we want to calculate the altitude (angle above the horizon) of a star in the night sky to determine our position on Earth. We can use the following expression:

$$\sin(Altitude) = \sin(Latitude) \cdot \sin(Declination) + \cos(Latitude) \cdot \cos(Declination) \cdot \cos(Hour\ Angle)]$$

Where:

$(Altitude)$ represents the altitude angle of the star.

$(Latitude)$ represents the observer's latitude on Earth.

$(Declination)$ represents the star's declination (celestial equivalent of latitude).

$(Hour\ Angle)$ represents the star's hour angle.

This formula, based on spherical trigonometry, allows navigators and astronomers to determine the altitude of celestial objects (e.g., stars, planets) using their observed declination and the observer's latitude. It's a crucial tool for celestial navigation at sea and for astronomical observations (Kessler, 2007; Kananizadeh & Momeni, 2017).

3.4.3. Engineering and Vibrations

Engineers use sine and cosine to design structures that withstand dynamic forces and vibrations. Engineers ensure the stability and integrity of bridges, buildings, and mechanical systems by analyzing oscillatory patterns using these functions. Vibrations in machinery, vehicles, and structures are studied using trigonometry to ensure optimal performance and safety (Fu et al., 2018).

Consider a mechanical system undergoing simple harmonic motion (SHM), such as a mass-spring-damper system. The displacement $x(t)$ of the mass m at time t can be described by the following equation:

$$x(t) = A\cos(\omega t)$$

Where:

$x(t)'$ represents the displacement of the mass as a function of time 't.'

A is the amplitude of the oscillation, indicating the maximum displacement from the equilibrium position.

ω is the angular frequency, defined as $(\omega = 2\pi f)$, where 'f' is the frequency of vibration in hertz (Hz).

t represents time.

This expression models the motion of the mass in SHM, showing how its displacement varies sinusoidally with time. Such equations are fundamental in engineering and physics for analyzing and designing systems subject to vibrations, such as mechanical structures, electrical circuits, and more.

3.4.4. Electrical Engineering and Signals

In electrical engineering, sine and cosine play a pivotal role in analyzing and designing electrical circuits and signals. These functions facilitate the analysis of alternating current (AC) circuits, helping engineers calculate voltage, current, and phase relationships. Signal processing techniques, including Fourier analysis, use sine and cosine to decompose complex signals into simpler components, revealing hidden patterns and frequencies (Kar et al., 2021).

Consider an alternating current (AC) voltage signal $V(t)$. It can be represented as follows:

$$V(t) = V_m \cdot \cos(2\pi ft + \phi)$$

Where:

$V(t)$ represents the instantaneous voltage at the time (t).

V_m is the peak voltage (maximum value) of the AC signal.

$\cos 2\pi ft + \phi$ denotes the cosine function.

2π represents the constant for converting between radians and cycles.

f is the frequency of the AC signal in hertz (Hz).

t is time.

ϕ is the phase angle, which determines the phase shift of the signal.

This expression describes an AC voltage signal $V(t)$ in terms of its peak voltage, frequency, and phase angle. It's a fundamental representation used in electrical engineering for analyzing and designing AC circuits and signals.

3.5. TRIGONOMETRIC IDENTITIES AND EQUATIONS

From geometry to music, navigation to engineering, and beyond, sine and cosine functions are used in countless applications. (Edmunds et al., 2012).

The following steps should be followed to solve a problem:

i. Examine the non-right angle present in the given problem.
ii. Identify concerned sides, including the adjacent, opposite, and hypotenuse.
iii. Identify which trigonometric ratio is determined by the two sides.
iv. Formulate an equation by commencing with the trigonometric ratio corresponding to the relevant angle and concluding with the division of two magnitudes.
v. Solve the equation to determine the desired outcome.

Here are some examples which use the above strategy (Kim, 2007):

Example: Find the length (denoted as x cm) in the right-angled triangle depicted in Figure 3.6.

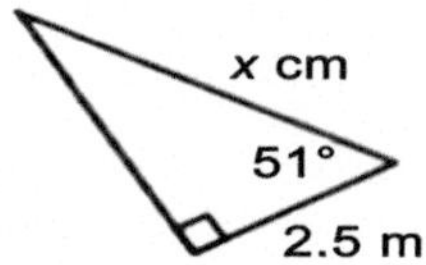

Figure 3.6. Right-angled triangle x cm.

Source: Michel, Creative Commons License.

The side measuring 2.5 cm is considered the neighboring side, while the side labeled as x cm represents the hypotenuse. The ratio of adjacent to hypotenuse becomes the cosine function (Shelupsky, 1959).

$$cos51° = \frac{2.5}{x}$$

Then, solve this equation for x.

$$x = 2.5 / cos51° = 3.972.$$

Example: In Figure 3.7, the length of each leg of a step ladder is 2.5 m. When the legs are opened out, the distance between their feet is 2 m. To find the angle between the legs, let AB and AC be the legs of the ladder. As there is no right angle, you have to make one by dropping the perpendicular AO from A to the base BC. The triangle ABC is isosceles, so AO bisects the angle BAC and the base BC.

Therefore,

$$BO = OC = 1m$$

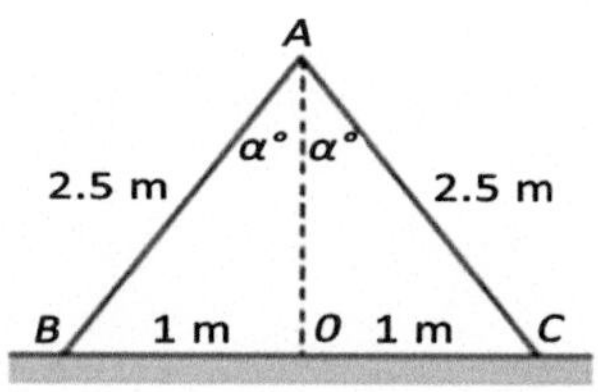

Figure 3.7. Triangle ABC.

Source: Daniel, Creative Commons License.

You need to find the angle BAC. Call it 2α°, so angle BAO = α°. The sides of length 1 m and 2.5 m are the opposite, and the hypotenuse for the angle α°, so you need the sine ratio.

Then

$$sin\alpha° = \frac{1}{2.5} = 0.4$$

So $\alpha° = 23.578$

And $2a° = 47.156...$

Therefore, the angle between the legs is 47.16° approximately (Radwan & Elwakil, 2012).

Example: A 30 m long ladder on a fire engine has to reach a window present at a height of 26 m from the ground. What angle, to the nearest degree, must it make with the ground, and how far from the building must it be placed? Let the ladder be AP (Figure 3.8), let θ° be the angle that the ladder makes with the ground and let d meters be the distance of the foot of the ladder from the window.

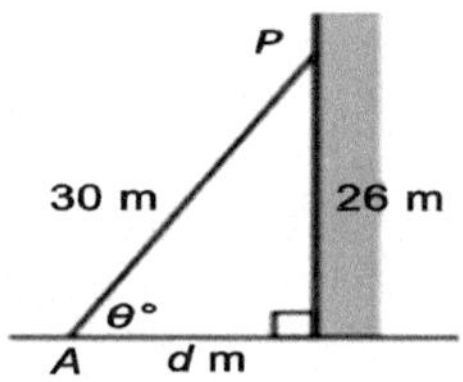

Figure 3.8. Ladder with angle AP.

Source: Daniel, Creative Commons License.

As the sides 26 m and 30 m are the opposite and the hypotenuse for the angle θ°, you need the sine ratio.

Then

$$sin\theta° = \frac{26}{30} = 0.8666$$

So

$$\theta° = 60.073$$

The ladder is placed at an angle of 60° to the ground. To find the distance d, it is best to use Pythagoras theorem.

$$d^2 = 30^2 - 26^2 = 900 - 676 = 224$$

So

$$d = 14.966$$

Example: The height of a cone is 18 cm, and the angle at the vertex is 88°. Find the slant height.

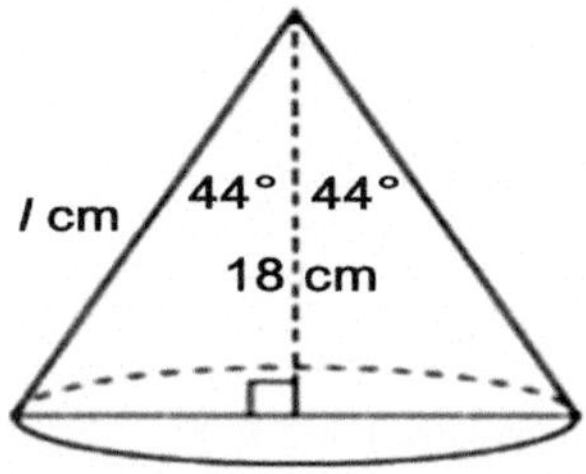

Figure 3.9. The cone with a height of 18cm.

Source: D.Turner, Creative Commons License.

In Figure 3.9, let "*l*" represent the measurement of the slant height of the cone, expressed in centimeters. Since the perpendicular to the base bisects the vertical angle of the cone, each part is 44°. The sides 18 cm and *l* cm are the adjacent and hypotenuse for the angle 44°, so the ratio concerned is the cosine function (Kantabutra, 1996).

Then

$$cos44° = \frac{18}{l}, l = \frac{18}{cos44°}$$

$$= \frac{18}{0.71933}$$

$$l = 25.022\,cm$$

The slant height is approximately 25.0 cm.

3.6. TRIGONOMETRIC RATIOS OF 45°, 30°& 60°

Trigonometric ratios for some elementary angles can be computed precisely.

3.6.1. Sine, Cosine and Tangent of 45°

Figure 3.10 depicts an isosceles right-angled triangle in which the two congruent sides measure 1 unit each.

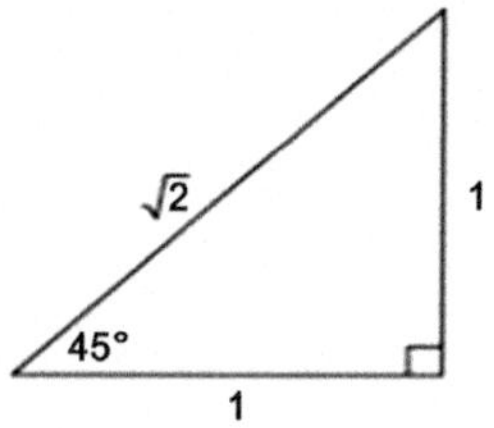

Figure 3.10. Right-angled triangle.

Source: Victor, Creative Commons License.

By applying Pythagoras's theorem, it can be determined that the hypotenuse length is equal to √2 units.

Hence, the trigonometric ratios corresponding to an angle of 45° are expressed as follows (Levy, 2015):

$$\sin 45° = \frac{1}{\sqrt{2}},\ \cos 45° = \frac{1}{\sqrt{2}}, tan\, 45° = 1$$

Using the equivalent values

$$sin45° = \frac{\sqrt{2}}{2}, cos45° = \frac{\sqrt{2}}{2}, tan45° = 1$$

Or,

$$\frac{1}{\sqrt{2}} = \frac{1}{\sqrt{2}} x \frac{\sqrt{2}}{\sqrt{2}} = \frac{\sqrt{2}}{2}.$$

3.6.2. Sine, Cosine and Tangent of 30°& 60°

Figure 3.11 shows an equilateral triangle with a side length of 2 units.

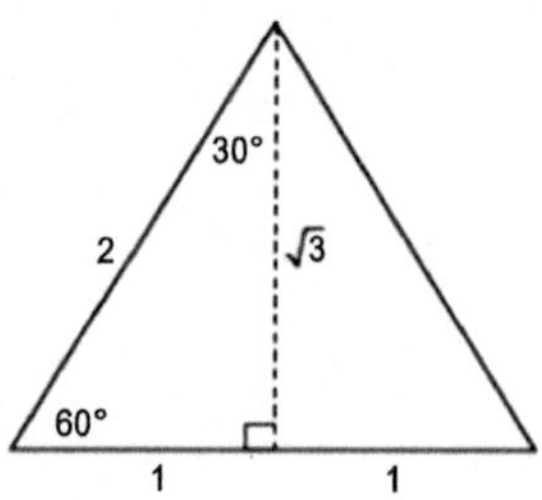

Figure 3.11. Equilateral triangle.

Source: Smith, Creative Commons License.

The perpendicular from the vertex bisects the base, dividing the original triangle into two triangles with angles of 30°, 60° and 90° and sides of lengths: 1 unit, 2 units and, using Pythagoras theorem, √3 units. Therefore, the trigonometric ratios for 30° are provided while the same ratios for 60° are given by:

$$sin30° = \frac{1}{2}, cos\,30° = \frac{\sqrt{3}}{2}, tan\,30° = \frac{1}{\sqrt{3}}\,or\,\frac{\sqrt{3}}{3}.$$

The same ratios for 60° are (Delice & Roper, 2006):

$$sin\,60° = \frac{\sqrt{3}}{2}, cos\,60° = \frac{1}{2}, tan60° = \sqrt{3}$$

3.6.3. Using the Calculator Accurately

When employing a calculator to solve an equation of the form sinθ°=23/37, it is imperative to strive for the highest possible level of accuracy in obtaining the solution.

Wrong Method:

$$sin\theta° = \frac{23}{37} = 0.622,$$

$$\theta° = 38.46$$

Correct Method:

$$sin\theta° = \frac{23}{37} = 0.6216216216$$

$$\theta° = 38.43$$

You can avoid the error by not writing down the three-significant figure approximation and by using the calculator in the following way.

$$sin\theta° = \frac{23}{37}$$

$$\theta° = 38.43$$

In this version, the answer to 23/37 was used directly to calculate the angle, and therefore, all the figures were preserved in the process. Sometimes, it may be necessary to use a calculator memory to store an intermediate answer to as many figures as you need. It is not necessary in this case, but you could calculate 23/37 and put the result into memory A. Then, you can calculate $\sin^{-1} A$ to get an accurate answer (Bhuria & Muralidhar, 2010).

3.7. SLOPE & GRADIENT

Figure 3.12 represents a lateral view of the section of a rising line segment (AC). AB is horizontal, and BC is the vertical line.

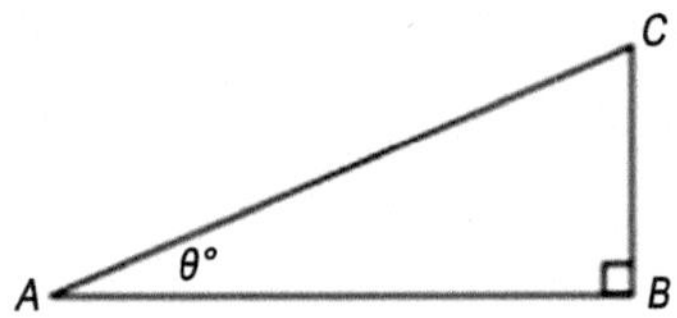

Figure 3.12. Side view of AC path.

Source: Plato, Creative Commons License.

Let θ° represent the angle formed between the rising line segment and the horizontal. The angle θ° is commonly called the slope angle or the path slope. The term "gradient of the path" refers to the ratio of tan θ°. In certain instances, particularly near railway tracks, the inclination is expressed as a ratio of 1 to 55 (Zhao & Burge, 2007).

$$gradient = \frac{1}{55}$$

When the angle of slope θ° is very small, as in the case of a railway and most roads, it makes little practical difference if you take sin θ° instead of tan θ° as the gradient. Also, in practice, it is easier to measure sin θ° (by measuring BC and AC), and the difference between AC and AB is relatively small, provided that θ° is small. You can use your calculator to see just how small the difference is between sines and tangents for small angles (Yu et al., 2022).

3.8. PROJECTIONS

In figure 3.13, let *l* be a straight line and let AB be a straight line segment, which makes an angle θ° with *l*.

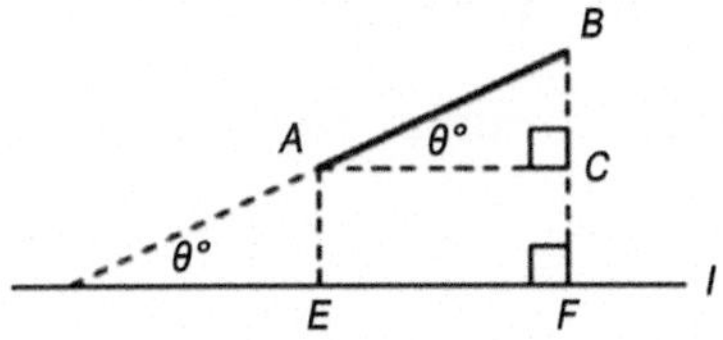

Figure 3.13. Figure explaining projections.

Perpendiculars are drawn from A and B to *l*, meeting *l* at E and F. Then EF is called the projection of AB on *l*. You can see from Figure 3.13 that the lengths AC and EF are equal.

As

$$\cos\theta° = \frac{AC}{AB}$$

So

$$AC = AB\cos\theta°$$

$$EF = AB\cos\theta°$$

3.9. HALF AND DOUBLE ANGLE FORMULAS

Half and double-angle formulas for sine and cosine are explained below:

3.9.1. Half-Angle Formulas

$$Sin\frac{A}{2} = \pm\sqrt{1-\cos A/2}$$

i. If A/2 is in the first or second quadrants, the formula uses the positive sign.
ii. If A/2 is in the third or fourth quadrants, the formula uses the negative sign.

$$Cos\frac{A}{2} = \pm\sqrt{1+\cos A/2}$$

iii. If A/2 is in the first or fourth quadrants, the formula uses the positive sign.
iv. If A/2 is in the second or third quadrants, the formula uses the negative sign (Dampil, 2014).

3.9.2. Double and Triple Angle Formulas

The double and triple-angle formulas are given below:

i. $Sin\,2A = 2SinA\,CosA$

ii. $Cos2A = Cos^2 A - Sin^2 A = 2Cos^2 A - 1 = 1 - Sin^2 A$

iii. $Sin\,3A = 3SinA - 4Sin^3 A$

iv. $Cos\,3A = 4Cos^3 A - 3CosA$

v. $Sin^2 A = 1 - Cos\left(2A\right)/2$

vi. $Cos^2 A = 1 + Cos\,(2A)/2$

3.10. CALCULUS & DERIVATIVE OF SINE AND COSINE

As we move further, we encounter the seamless integration of the sine and cosine functions with the elegant concepts of calculus. The derivative, a fundamental concept in calculus, reveals the intricate rates of change and slopes inherent to these trigonometric functions (Siyepu, 2015).

3.10.1. The Derivative

The derivative of a function captures the instantaneous rate of change of that function at any given point. When applied to the sine and cosine functions, the derivative shows the nuances of their oscillations and the variations in their slopes as angles shift (Püschel & Moura, 2003).

For the sine function, sin(θ), its derivative results in the cosine function, cos(θ).

$$\frac{d}{d\theta} sin\left(\theta\right) = cos\left(\theta\right)$$

Conversely, when we examine the derivative of the cosine function (cos(θ)), it reveals the negative of the sine function, providing a compelling reflection of their alternating behavior (Wazwaz, 2006):

$$\frac{d}{d\theta} cos\left(\theta\right) = -sin\left(\theta\right)$$

3.10.2. Analyzing Rates of Change

The derivative of sine and cosine offers insights into these functions' oscillatory nature and dynamic behavior. These derivatives encapsulate the instantaneous slopes of the sine and cosine functions, offering a deeper understanding of how they evolve as angles vary.

The fundamental relationship between the original functions and their derivatives is a cornerstone for solving intricate problems across mathematics, physics, and engineering. Whether modeling wave behavior, analyzing harmonic motion, or predicting dynamic systems, the derivatives of sine and cosine play a pivotal role in capturing the essence of changing quantities (Thompson, 2008).

3.10.3. Applications

The calculus of sine and cosine finds applications in diverse fields, ranging from physics and engineering to signal processing and mathematics. Engineers leverage the derivatives of these functions to model vibrations, analyze waveforms, and design systems with dynamic components. In physics, the derivatives help describe the behavior of oscillatory motion, wave propagation, and ACs. The derivatives' significance extends to solving differential equations that govern various phenomena, highlighting their pervasive influence (Ninness et al., 2009).

3.11. COMPLEX NUMBERS

Exploring complex numbers unveils a realm where imagination meets mathematics, where real and imaginary components are explored in detail. The sine and cosine functions take on new dimensions, intertwining with complex numbers to create a symphony of intricate relationships (Mathews & Mathews, 1999).

Complex numbers, denoted as $z = a + bi$, comprise a real part (*a*) and an imaginary part (*bi*), where (*i*-expressed as iota) is the imaginary unit ($i^2 = -1$). These numbers transcend the confines of the real number line, introducing a vertical dimension that unleashes new mathematical possibilities.

The foundation of complex numbers paves the way for Euler's Formula, a cornerstone of this exploration:

$$e^{ix} = cos(x) + isin(x)$$

This formula acts as a bridge, connecting the exponential function with the sine and cosine functions. It links the real and imaginary worlds, revealing a profound connection between the oscillatory behavior of trigonometric functions and the exponential growth of complex numbers (Sarwagya et al., 2020).

3.12. NOTATION

The notation focuses on determining non-right triangles' side lengths and angle magnitudes. A notation system to represent the sides and angles of triangles is quite advantageous.

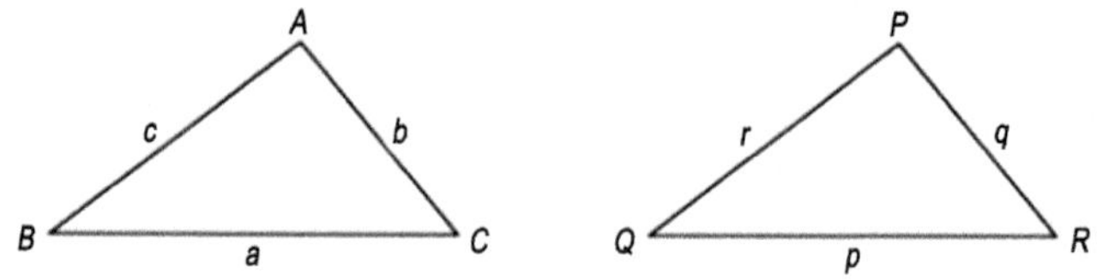

Figure 3.14. Two triangles, ABC and PQR.

Source: Smith, Creative Commons License.

In the triangle ABC, the angles are called A, B and C; the sides opposite to these angles are given the corresponding lower-case letters, a, b and c. In the triangle PQR, the sides opposite to the angles P, Q, and R are p, q and r. Here, the angles will be measured in degrees, so in figure 3.14, A= 95°.

The sides may be measured in any units you choose, provided they are all measured in the same units. In some of the diagrams that follow, the units of length are omitted (Koornwinder & Swarttouw, 1992).

3.12.1. Area of a Triangle

Finding the area of a triangle is a fundamental geometric principle that lies at the intersection of mathematics and real-world applications. Understanding how to calculate the area of a triangle is a key skill in geometry, helpful in measuring the space enclosed within its three sides. This seemingly simple concept carries profound implications, serving as a cornerstone in various fields, from architecture and engineering to geography and art.

To calculate the area of a triangle, one can employ several methods, each rooted in the relationships between its sides and angles. One of the most fundamental approaches is the use of base and height. Given the length of a base and its corresponding height, the area can be calculated as half the product of the base and height. This method is particularly useful for right triangles, where one of the sides is the base, and the perpendicular drawn from the vertex opposite to the base is the height (Nicol, 1993).

For more general triangles, the concept of trigonometry comes into play. The sine function enables us to relate the area of a triangle to the lengths of two sides and the sine of the angle between them. The formula: $Area = \frac{1}{2} \times base \times height,$ is expanded to $Area = \frac{1}{2} \times side_1 \times side_2 \times sin\theta,$ where θ is the angle between the two sides. This trigonometric approach provides a versatile means of calculating the area of triangles of varying lengths and angles.

In architecture and engineering, calculating the area of triangular structures is essential for designing stable and balanced frameworks. Cartographers rely on this concept to measure the surface area of land masses and create accurate maps. Artists and designers incorporate triangular compositions to create visually appealing and dynamic artworks.

When determining the area of a triangle, it is necessary to consider whether the triangle is acute-angled or obtuse-angled. To derive a formula for the area of a triangle about its sides and angles, it is necessary to analyze acute-angled and obtuse-angled triangles as distinct cases (Krantz, 1992).

- Case I: Acute-Angled Triangle:

Figure 3.15 shows an acute-angled triangle with the perpendicular drawn from A onto BC. Let the length of this perpendicular be *h*. Let the area of the triangle be Δ, so:

$$\Delta = \frac{1}{2}ab$$

To find *h* you can use the left-hand right-angled triangle.

Thus,

$$sin°B = \frac{h}{c}, so\, h = csinB°$$

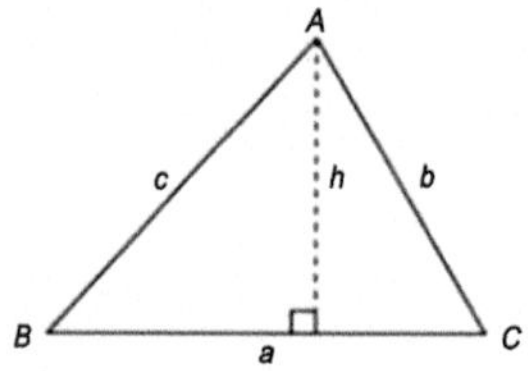

Figure 3.15. Acute angled triangle.

Source: Smith, Creative Commons License.

Therefore;

$$\Delta = \frac{1}{2}ah = \frac{1}{2}acsin\,B°$$

$\therefore h = csin\theta$ **Equation 1**

Alternatively, you could find h from the other right-angled triangle,

$$sinC° = \frac{h}{b} so\, h = bsin\,C°$$

So

$$\Delta = \frac{1}{2}ah = \frac{1}{2}absin\,C°$$

Equation 2

So;

$$\Delta = \frac{1}{2}acsin\,B° = \frac{1}{2}absin\,C°$$

Equation 3

Notice the symmetry of the two expressions in Equation 3. Each of the expressions $(\frac{1}{2}ac\sin B°$ and $\frac{1}{2}ab\sin C°)$ consists of ½ the product of two sides multiplied by the sine of the angle between them. It follows from this symmetry that there is a third expression for the area of the triangle, namely $\frac{1}{2}bc\sin A°$ You could have derived this expression directly if, in Figure 3.15, you had drawn the perpendicular from B to AC, or from C to AB. Thus, for an acute-angled triangle.

$$\Delta = \frac{1}{2}bcsin\,A° = \frac{1}{2}casin\,B° = \frac{1}{2}absinsin\,C°$$

- Case II: Obtuse Angled Triangle

Figure 3.16 shows a triangle, obtuse angled at C, with the perpendicular drawn from A onto BC. Let the length of this perpendicular be *h*.

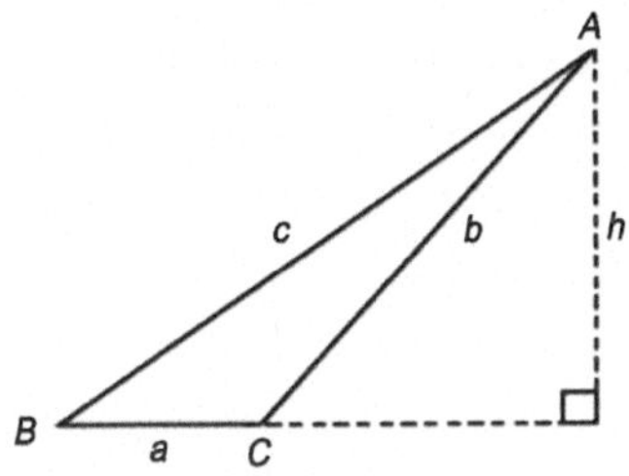

Figure 3.16. Obtuse triangle.

Source: Dan, Creative Commons License.

Let the area of the triangle be Δ, so $\Delta = \frac{1}{2}ah$.

Thus,

$$sinB° = \frac{h}{c}, so\, h = csinB°$$

Therefore,

$$\Delta = \frac{1}{2}ah = \frac{1}{2}ac\, sinB°$$ **Equation 4**

Alternatively,

$$sin(180° - C°) = \frac{h}{b} so\, h = bsin(180° - C°)$$

So

$$\Delta = \frac{1}{2}ah = \frac{1}{2}absin(180° - C°)$$ **Equation 5**

And;

$$sinC° = sin(180° - C°)$$

Therefore, Equation 5 can be written as:

$$\Delta = \frac{1}{2}absin(180° - C°) = \frac{1}{2}absin\,C°.$$

Putting Equations 4 and 5 together,

$$\Delta = \frac{1}{2} ac \sin B° = \frac{1}{2} ab \sin C°$$ **Equation 6**

The symmetrical nature of the two formulas in Equation 6 is evident again. Both formulas, 1/2 ac sin(B°) and 1/2 ab sin(C°), represent the product of half the multiplication of two sides and the sine of the angle formed between them (Harrison, 2002).

Similarly,

$$\frac{1}{2} ab \sin A° .$$

Thus, for an obtuse-angled triangle,

$$\Delta = \frac{1}{2} ab \sin A° = \frac{1}{2} ca \sin B° = \frac{1}{2} ab \sin C°$$

- **Case 3: Area of Any Triangle:**

Thus, the area Δ of any triangle, acute-angled or obtuse-angled, is

$$\Delta = \frac{1}{2} bc \sin A° = \frac{1}{2} ca \sin B° = \frac{1}{2} ab \sin C°$$

Or,

$$Area = \frac{1}{2} \times product\ of\ sides \times sine\ of\ included\ angle$$

3.12.2. The Sine Formula for a Triangle

The formula for the area is:

$$\Delta = \frac{1}{2} bc \sin A° = \frac{1}{2} ca \sin B° = \frac{1}{2} ab \sin C°$$

Or,

$$bc \sin A° = ca \sin B° = ab \sin C°$$

Or,

$$\sin A° / a = \sin B° / b = \sin C° / c .$$

This formula is called the ***sine formula for a triangle*** or, more briefly, the ***sine formula.***

3.12.3. The Cosine Formula for a Triangle

The sine formula is easy to remember and easy to use, but it is no help if you know the lengths of two sides and the included angle and wish to find the length of the third side. In Figure 3.17, suppose that you know the lengths of sides a and b, and the angle C between them and that you wish to find the length of the side c (Gervois & Navelet, 1984).

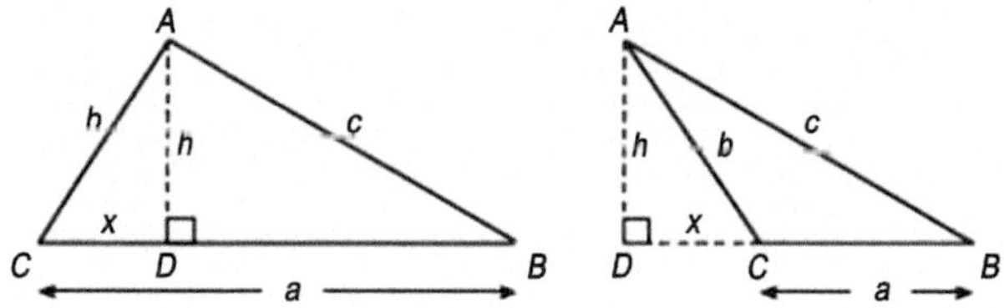

Figure 3.17. Triangle ABC.

Source: Dan, Creative Commons License.

There are two cases to consider: when angle C is acute and when angle C is obtuse. In both cases, the perpendicular from A is drawn, meeting BC at D. The length of this perpendicular is *h*. Let the length CD be x (Lukarevski, 2023).

- Case 1: Acute Angled Triangle:

In the triangle ACD (Figure 3.17):

$$h^2 = b^2 - x^2$$

Using Pythagoras's theorem in the triangle ABD:

$$h^2 = c^2 - (a-x)^2$$

Equating for h^2 we get:

$$b^2 - x^2 = c^2 - (a-x)^2$$

That is

$$b^2 - x^2 = c^2 - a^2 + 2ax - x^2$$

Or

$$c^2 = a^2 + b^2 - 2ax$$

In triangle ACD, x = b cos C°, so the expression for c^2 becomes

$$c^2 = a^2 + b^2 - 2abcos\,C°$$

- Case 2: Obtuse-Angled Triangle

In the triangle ABD (Figure 3.17):

$$h^2 = b^2 - x^2$$

Using Pythagoras's theorem in the triangle ABD:

$$h^2 = c^2 - (a + x)^2$$

Equating for h^2

$$b^2 - x^2 = c^2 - (a + x)^2$$

That is

$$b^2 - x^2 = c^2 - a^2 + 2ax - x^2$$

Or

$$c^2 = a^2 + b^2 + 2ax$$

In triangle ACD, x = b cos (180° − C°). However, from the graph of y = cos θ°, cos(180° − C°) = −cos C° (Sigler et al., 2017).

Thus, when C° is obtuse, x = −b cos C°. The expression for c^2 therefore becomes:

$$c^2 = a^2 + b^2 - 2abcoscos\,C°$$

This formula is called the *cosine formula for a triangle* or, more briefly, the *cosine formula.*

3.13. ADVANCED CONCEPTS AND CURRENT RESEARCH

Trigonometry, often recognized by its sine and cosine functions, extends its reach beyond these foundational elements into advanced concepts and pioneering research. This section will explain the intricate relationships, abstract frameworks, and interdisciplinary applications, shedding light on the forefront of trigonometric exploration that transcends the boundaries of traditional understanding (Khafaga et al., 2022).

3.13.1. Non-Euclidean Geometries and Trigonometry

Exploring non-Euclidean geometries introduces a captivating relationship between trigonometry and the curvature of space. Hyperbolic and elliptic geometries introduce novel perspectives to familiar trigonometric relationships. Hyperbolic trigonometry, for instance, delves into triangles on surfaces with constant negative curvature, revealing alternative interpretations of sine, cosine, and tangent functions. These advanced geometries challenge conventional notions of angles and distances, reshaping the application of trigonometry in unconventional spatial landscapes (Rosenfeld, 2012).

3.13.2. Fourier Analysis and Signal Processing

The key to unlocking the intricate symphonies concealed within complex signals and functions lies within trigonometry. Fourier analysis, an advanced concept rooted in trigonometry, facilitates the decomposition of functions into elemental trigonometric components. This process unveils the underlying frequencies and patterns embedded in real-world phenomena (Razaq et al., 2022).

3.13.3. Current Research Frontiers

Mathematicians are diligently exploring innovative methodologies for solving intricate trigonometric equations and expressing trigonometric functions in terms of other mathematical constructs. Leveraging the advancements in symbolic computation and computer algebra systems, researchers are equipped to tackle complex trigonometric challenges using computational tools. Furthermore, interdisciplinary collaborations bridge trigonometry with computer graphics, artificial intelligence, and machine learning, showcasing its vital role in shaping the trajectory of modern technologies (Zhou et al., 2021).

3.14. SUMMARY

Exploring the chapter on sine and cosine in trigonometry has been a captivating journey into the heart of these fundamental functions. Sine and cosine, with their rhythmic oscillations and geometric underpinnings, have formulated an intricate amalgam of knowledge that transcends disciplinary boundaries.

At the outset, the definitions of sine and cosine came to the forefront, revealing their deep-rooted connection to right triangles and the unit circle. Sine, symbolized as sin θ, emerged as the vertical projection of an angle onto the unit circle, while cosine, cos θ, represented the horizontal projection. These definitions provided a fundamental toolkit for measuring angles, distances, and periodic phenomena, forming the bedrock for their applications in various fields.

The chapter unveiled the special trigonometric ratios associated with 45, 30, and 60 degrees. These angles, known for their distinct geometric properties, allowed us to solve practical problems involving angles of elevation, depression, and inclination. By understanding the sine, cosine, and tangent ratios for these special angles, insights into the geometric relationships that underlie trigonometry's foundations are gained.

In the context of slopes and gradients, sine and cosine functions provided a powerful means to quantify inclines and declines. The concept of projection was elegantly demonstrated through the orthogonal projection of a vector onto a line, showcasing how sine and cosine facilitate the decomposition of complex quantities into simpler components—a fundamental concept with widespread applications.

Exploring sine and cosine formulas illuminated the rich tapestry of identities that these functions weave. The Pythagorean identity, $\sin^2(\theta) + \cos^2(\theta) = 1$, stood out as a foundational relationship that captured the interplay between sine and cosine. These formulas provided a versatile toolkit for simplifying expressions, solving equations, and constructing elegant proofs—a testament to the enduring power of trigonometric identities.

Calculus offered a deeper perspective on sine and cosine, unveiling their connection to the derivative. The cyclical nature emerged—the derivative of sine is cosine, and the derivative is the negative sine. This profound relationship underscored the integral role of sine and cosine in studying motion, waves, and change, emphasizing their presence as essential tools in the mathematical toolbox.

Complex numbers were introduced to Euler's formula, a striking bridge between algebra and geometry. This formula, $e^{i\theta} = \cos(\theta) + i\sin(\theta)$ demonstrated the intricate relationship between exponential, sine, and cosine functions. Beyond its elegance, Euler's formula found applications in physics, engineering, and complex analysis, reinforcing the pervasive reach of sine and cosine.

The significance of notation and conventions in trigonometry became apparent, ensuring clarity and precision in mathematical discourse. The application of sine and cosine in determining the area of triangles illustrated their geometric significance—a concept inherently linked to the fundamental principles of trigonometry.

REFERENCES

1. Abualigah, L., & Diabat, A., (2021). Advances in sine cosine algorithm: A comprehensive survey. *Artificial Intelligence Review, 54*(4), 2567–2608.
2. Akkılıc, A. N., Sulaiman, T. A., & Bulut, H., (2021). Applications of the extended rational sine-cosine and Sinh-Cosh techniques to some nonlinear complex models arising in mathematical physics. *Applied Mathematics and Nonlinear Sciences, 6*(2), 19–30.
3. Al-Mohy, A. H., Higham, N. J., & Relton, S. D., (2015). New algorithms for computing the matrix sine and cosine separately or simultaneously. *SIAM Journal on Scientific Computing, 37*(1), A456-A487.
4. Alotaibi, A., Arif, M., Alghamdi, M. A., & Hussain, S., (2020). Starlikness associated with cosine hyperbolic function. *Mathematics, 8*(7), 1118.
5. Arel, I., Rose, D. C., & Karnowski, T. P., (2010). Deep machine learning-a new frontier in artificial intelligence research [research frontier]. *IEEE Computational Intelligence Magazine, 5*(4), 13–18.
6. Bag, A. K., (1976). Madhava's sine and cosine series. *Indian Journal of History of Science, 11*(1), 54–57.
7. Bass, L., Su, T., Chesnavich, W. J., & Bowers, M. T., (1975). Ion-polar molecule collisions. A modification of the average dipole orientation theory: The cos θ model. *Chemical Physics Letters, 34*(1), 119–122.
8. Berlinghoff, W. P., & Gouvêa, F. Q., (2021). *Math Through the Ages: A Gentle History for Teachers and Others Expanded* (2nd edn., Vol. 32, pp. 85–86). American Mathematical Soc.
9. Bhuria, S., & Muralidhar, P., (2010). FPGA implementation of sine and cosine value generators using CORDIC algorithm for satellite attitude determination and calculators. In: *2010 International Conference on Power, Control and Embedded Systems* (Vol. 1, pp. 1–5). IEEE.
10. Chesneau, C., Bakouch, H. S., & Hussain, T., (2019). A new class of probability distributions via cosine and sine functions with applications. *Communications in Statistics-Simulation and Computation, 48*(8), 2287–2300.
11. Cooley, J. W., Lewis, P. A. W., & Welch, P., (1970). The fast Fourier transform algorithm: Programming considerations in the calculation of sine, cosine, and Laplace transforms. *Journal of Sound and Vibration, 12*(3), 315–337.

12. Dampil, F. G., (2014). Development of trigonometric formula for Sin4n [theta] and Cos4n [theta] using half angle identities as mathematical algorithm. *International Journal of Applied Physics and Mathematics, 4*(3), 155.
13. Das, P., Juyal, A., & Moody, D., (2017). Integral isosceles triangle–parallelogram and heron triangle–rhombus pairs with a common area and common perimeter. *Journal of Number Theory, 180*, 208–218.
14. Delice, A., & Roper, T., (2006). Implications of a comparative study for mathematics education in the English education system. *Teaching Mathematics and Its Applications: International Journal of the IMA, 25*(2), 64–72.
15. Edmunds, D. E., Gurka, P., & Lang, J., (2012). Properties of generalized trigonometric functions. *Journal of Approximation Theory, 164*(1), 47–56.
16. Ehteram, M., Panahi, F., Ahmed, A. N., Mosavi, A. H., & El-Shafie, A., (2022). Inclusive multiple model using hybrid artificial neural networks for predicting evaporation. *Frontiers in Environmental Science, 9*, 789995.
17. Fauvel, J., (1995). Revisiting the history of logarithms. *Learn from the Masters, 74*, 39–48.
18. Fletcher, A., (1955). Table of sine and cosine integrals for arguments from 10 to 100. Pp. xv, 187. $2.25. 1954. Applied mathematics series, 32. (National Bureau of Standards, Washington). *The Mathematical Gazette, 39*(330), 350, 351.
19. Fu, W., Wang, K., Li, C., Li, X., Li, Y., & Zhong, H., (2018). Vibration trend measurement for a hydropower generator based on optimal variational mode decomposition and an LSSVM improved with chaotic sine cosine algorithm optimization. *Measurement Science and Technology, 30*(1), 015012.
20. Garcia, E. O., Cumplido, R., & Arias, M., (2006). Pipelined CORDIC design on FPGA for a digital sine and cosine waves generator. In: *2006 3rd International Conference on Electrical and Electronics Engineering* (Vol. 1, pp. 1–4). IEEE.
21. Garfunkel, J., (1969). Exploring geometric maxima and minima. *The Mathematics Teacher, 62*(2), 85–90.
22. Gervois, A., & Navelet, H., (1984). Some integrals involving three Bessel functions when their arguments satisfy the triangle inequalities.

Journal of Mathematical Physics, 25(11), 3350–3356.

23. Glaisher, J. W. L., (1870). XX. Tables of the numerical values of the sine-integral, cosine-integral, and exponential-integral. *Philosophical Transactions of the Royal Society of London*, (160), 367–388.
24. Greenwood, J., (2002). The correct and incorrect generation of a cosine distribution of scattered particles for Monte-Carlo modelling of vacuum systems. *Vacuum, 67*(2), 217–222.
25. Harrison, E. P., (2002). Sharing teaching ideas: Using the law of cosines to teach the ambiguous case of the law of sines. *The Mathematics Teacher, 95*(2), 114–116.
26. Ibrahim, K. M., & Abdul-Karim, M. A., (1986). Digital impedance measurement based on inverse sine function module. *IEEE Transactions on Instrumentation and Measurement*, (1), 87, 88.
27. Jackson, F. H., (1903). A basic-sine and cosine with symbolical solutions of certain differential equations. *Proceedings of the Edinburgh Mathematical Society, 22*, 28–39.
28. Kalmykov, M. Y., & Sheplyakov, A., (2005). lsjk—a C++ library for arbitrary-precision numeric evaluation of the generalized log-sine functions. *Computer Physics Communications, 172*(1), 45–59.
29. Kananizadeh, R., & Momeni, O., (2017). High-power and high-efficiency millimeter-wave harmonic oscillator design, exploiting harmonic positive feedback in CMOS. *IEEE Transactions on Microwave Theory and Techniques, 65*(10), 3922–3936.
30. Kantabutra, V., (1996). On hardware for computing exponential and trigonometric functions. *IEEE Transactions on Computers, 45*(3), 328–339.
31. Kar, M. K., Kumar, S., Singh, A. K., & Panigrahi, S., (2021). A modified sine cosine algorithm with ensemble search agent updating schemes for small signal stability analysis. *International Transactions on Electrical Energy Systems, 31*(11), e13058.
32. Katz, V. J., (1986). Using history in teaching mathematics. *For the Learning of Mathematics, 6*(3), 13–19.
33. Kaur, M., Kaur, R., Singh, N., & Dhiman, G., (2021). SChoA: A newly fusion of sine and cosine with chimp optimization algorithm for HLS of datapaths in digital filters and engineering applications. *Engineering with Computers, 98*(2), 1–29.
34. Kawrakow, I., (2000). Accurate condensed history Monte Carlo

simulation of electron transport. I. EGSnrc, the new EGS4 version. *Medical Physics, 27*(3), 485–498.

35. Kessler, B., (2007). A "sound" approach to Fourier transforms: Using music to teach trigonometry. In: *2007 Bridges Donostia Conference Proceedings* (Vol. 2, pp. 135–136).
36. Khafaga, D. S., Alhussan, A. A., El-Kenawy, E. S. M., Ibrahim, A., Eid, M. M., & Abdelhamid, A. A., (2022). Solving optimization problems of metamaterial and double T-shape antennas using advanced meta-heuristics algorithms. *IEEE Access, 10*, 74449–74471.
37. Kim, T., (2007). q-extension of the Euler formula and trigonometric functions. *Russian Journal of Mathematical Physics, 14*(3), 275–278.
38. Koornwinder, T. H., & Swarttouw, R. F., (1992). On *q*-analogues of the Fourier and Hankel transforms. *Transactions of the American Mathematical Society, 333*(1), 445–461.
39. Krantz, S. G., (1992). On the area inside a circle. *Missouri Journal of Mathematical Sciences, 4*(1), 2–8.
40. Leindler, L., (2002). A new class of numerical sequences and its applications to sine and cosine series. *Analysis Mathematica, 28*(4), 279–286.
41. Levy, S. Y., (2015). This paper develops the digit-by-digit algorithms including evaluation of sine, cosine, arctangent, and the hyperbolic functions of Sinh. *Computer Arithmetic, 1*, 261.
42. Liu, Z., Xu, L., Liu, T., Chen, H., Li, P., Lin, C., & Liu, S., (2011). Color image encryption by using Arnold transform and color-blend operation in discrete cosine transform domains. *Optics Communications, 284*(1), 123–128.
43. Lukarevski, M., (2023). A new look at the fundamental triangle inequality. *Mathematics Magazine, 96*(2), 141–149.
44. Mahmud, S. M., (1990). High precision phase measurement using reduced sine and cosine tables. *IEEE Transactions on Instrumentation and Measurement, 39*(1), 56–60.
45. Martín-Fernández, E., Ruiz-Hidalgo, J. F., & Rico, L., (2019). Meaning and understanding of school mathematical concepts by secondary students: The study of sine and cosine. *Eurasia Journal of Mathematics, Science, and Technology Education, 15*(12), 3698–3699.
46. Martucci, S. A., (1994). Symmetric convolution and the discrete sine

and cosine transforms. *IEEE Transactions on Signal Processing, 42*(5), 1038–1051.

47. Mathews, S., & Mathews, K., (1999). Some enrichment ideas for complex algebra in the college prep curriculum. *Problems, Resources, and Issues in Mathematics Undergraduate Studies, 9*(3), 251–261.
48. Neugebauer, O., (2012). *A History of Ancient Mathematical Astronomy* (Vol. 1, pp. 325–328). Springer Science & Business Media.
49. Nicol, S. J., (1993). Integrals of products of sine and cosine with different arguments. *The College Mathematics Journal, 24*(2), 158–160.
50. Nieto, M. M., (1993). Quantum phase and quantum phase operators: Some physics and some history. *Physica Scripta, 1993*(48), 574.
51. Ninness, C., Dixon, M., Barnes-Holmes, D., Rehfeldt, R. A., Rumph, R., McCuller, G., & McGinty, J., (2009). Constructing and deriving reciprocal trigonometric relations: A functional analytic approach. *Journal of Applied Behavior Analysis, 42*(2), 191–208.
52. Peirce, B., (1845). *An Elementary Treatise on Plane & Spherical Trigonometry: With Their Applications to Navigation, Surveying, Heights, and Distances, and Spherical Astronomy, and Particularly Adapted to Explaining the Construction of Bowditch's Navigator, and the Nautical Almanac* (Vol. 74, No. 34, pp. 98710–98713). J. Munroe.
53. Püschel, M., & Moura, J. M., (2003). The algebraic approach to the discrete cosine and sine transforms and their fast algorithms. *SIAM Journal on Computing, 32*(5), 1280–1316.
54. Radwan, A. G., & Elwakil, A. S., (2012). An expression for the voltage response of a current-excited fractance device based on fractional-order trigonometric identities. *International Journal of Circuit Theory and Applications, 40*(5), 533–538.
55. Razaq, A., Ahmad, M., Yousaf, A., Alawida, M., Ullah, A., & Shuaib, U., (2022). A group theoretic construction of large number of AES-like substitution-boxes. *Wireless Personal Communications, 122*(3), 2057–2080.
56. Rosenfeld, B. A., (2012). *A History of Non-Euclidean Geometry: Evolution of the Concept of a Geometric Space* (Vol. 12, pp. 233–235). Springer Science & Business Media.
57. Sarwagya, K., Nayak, P. K., & Ranjan, S., (2020). Optimal coordination of directional overcurrent relays in complex distribution networks

using sine cosine algorithm. *Electric Power Systems Research, 187*, 106435.

58. Sharat, K., Uma, B. V., & Sagar, D. M., (2014). Calculation of Sine and cosine of an angle using the CORDIC algorithm. *International Journal of Innovative Technology and Research (IJITR),* 891–895.
59. Shelupsky, D., (1959). A generalization of the trigonometric functions. *The American Mathematical Monthly, 66*(10), 879–884.
60. Shiva, C. K., Vedik, B., Raj, S., Mahapatra, S., & Mukherjee, V., (2023). A novel chaotic chimp sine cosine algorithm part-II: Automatic generation control of complex power system. *Chaos, Solitons & Fractals, 173*, 113673.
61. Sigler, A., Stupel, M., & Flores, A., (2017). Relations among five radii of circles in a triangle, its sides, and other segments. *International Journal of Mathematical Education in Science and Technology, 48*(5), 782–793.
62. Siyepu, S. W., (2015). Analysis of errors in derivatives of trigonometric functions. *International Journal of STEM Education, 2*, 1–16.
63. Sneeuw, N., (1994). Global spherical harmonic analysis by least-squares and numerical quadrature methods in historical perspective. *Geophysical Journal International, 118*(3), 707–716.
64. Stolwijk, A. M., Straatman, H. M. P. M., & Zielhuis, G. A., (1999). Studying seasonality by using sine and cosine functions in regression analysis. *Journal of Epidemiology & Community Health, 53*(4), 235–238.
65. Takagi, N., Asada, T., & Yajima, S., (1991). Redundant CORDIC methods with a constant scale factor for sine and cosine computation. *IEEE Transactions on Computers, 40*(09), 989–995.
66. Tang, H., Murugusundaramoorthy, G., Li, S. H., & Ma, L. N., (2022). Fekete-Szegö and Hankel inequalities for certain class of analytic functions related to the sine function. *AIMS Math, 7*, 6365–6380.
67. Thompson, P. W., (2008). Conceptual analysis of mathematical ideas: Some spadework at the foundations of mathematics education. In: *Proceedings of the Annual Meeting of the International Group for the Psychology of Mathematics Education* (Vol. 1, pp. 31–49). Morelia: PME.
68. Wazwaz, A. M., (2006). The sine–cosine and the tanh methods: Reliable tools for analytic treatment of nonlinear dispersive equations. *Applied Mathematics and Computation, 173*(1), 150–164.

69. Xie, N., Smith, R. A., Mukhopadhyay, S., & Hallett, S. R., (2018). A numerical study on the influence of composite wrinkle defect geometry on compressive strength. *Materials & Design, 140*, 7–20.
70. Yang, J., Li, G., Zhang, Y., Wu, J., Zhang, M., Li, J., & Liu, W., (2022). Blind crosstalk correction for color fringe projection profilometry combining the sine and cosine phase-shifting patterns. *Journal of Sensors, 2022*, 1–12.
71. Yu, S., Chen, Z., Heidari, A. A., Zhou, W., Chen, H., & Xiao, L., (2022). Parameter identification of photovoltaic models using a sine cosine differential gradient-based optimizer. *IET Renewable Power Generation, 16*(8), 1535–1561.
72. Zhao, C., & Burge, J. H., (2007). Orthonormal vector polynomials in a unit circle, Part I: Basis set derived from gradients of Zernike polynomials. *Optics Express, 15*(26), 18014–18024.
73. Zhou, J., Qiu, Y., Armaghani, D. J., Zhang, W., Li, C., Zhu, S., & Tarinejad, R., (2021). Predicting TBM penetration rate in hard rock condition: A comparative study among six XGB-based metaheuristic techniques. *Geoscience Frontiers, 12*(3), 101091.

CHAPTER 4

Identities and Formulas

4.1. INTRODUCTION

The fundamental principles of trigonometry are built upon several identities and formulas that provide the mathematical framework for understanding and solving a wide range of problems in geometry, physics, engineering, and more.

One of the most fundamental concepts in trigonometry is the Pythagorean theorem, expressed as $\left(a^2+b^2=c^2\right)$, where (a) and (b) are the lengths of the two shorter sides of a right triangle, and (c) is the length of the longest side, known as the hypotenuse. This theorem establishes a critical relationship between the sides of a right triangle.

Trigonometry further extends its applications through sum and difference formulas, which allow for the manipulation and simplification of trigonometric expressions. These formulas, exemplified by $\sin(A \pm B)=\sin(A)\cos(B) \pm \cos(A)\sin(B)$, enable the break down of trigonometric functions of sums and differences into products of trigonometric functions, making complex calculations more manageable. They are particularly valuable in fields like physics and engineering, where analyzing oscillations, waves, and periodic phenomena is common.

Another set of essential formulas in trigonometry includes the double angle formulas, such as $\sin(2\theta)=2\sin(\theta)\cos(\theta)$ and $\cos(2\theta)=\cos^2(\theta)-\sin^2(\theta)$. These formulas provide relationships between trigonometric functions of double angles and single angles. They find applications in various fields, including wave interference patterns, optics, and signal processing.

The half-angle formulas, represented by $\left(\sin\left(\frac{\theta}{2}\right)=\pm\sqrt{\frac{1-\cos(\theta)}{2}}\right)$ and

$$\left(\cos\left(\frac{\theta}{2}\right)=\pm\sqrt{\frac{1+\cos(\theta)}{2}}\right),$$

are crucial when dealing with angles that are half their original size. These formulas find utility in solving problems involving rotation and symmetry, as well as in calculus.

The Law of Sines:\

$$\left(\frac{\sin(A)}{a}=\frac{\sin(B)}{b}=\frac{\sin(C)}{c}\right)$$

and the Law of Cosines: $c^2 = a^2 + b^2 - 2ab\cos(C)$, are foundational for solving triangles and understanding the relationships between their angles and sides. The Law of Sines is particularly useful for solving non-right triangles, providing a simple ratio between the sine of an angle and the length of the side opposite that angle. On the other hand, the Law of Cosines is a powerful tool for solving triangles when the lengths of all three sides or two sides and the included angle are known.

In fields like surveying, architecture, and navigation, trigonometric principles are used to measure distances, heights, and angles. In physics, these concepts help model waveforms, analyze the motion of objects, and understand the behavior of light and sound. In engineering, they are essential for designing structures, circuits, and control systems.

4.2. LIST OF TRIGONOMETRIC IDENTITIES

Trigonometry contains a variety of mathematical identities that are useful for solving a wide range of trigonometric problems. Using trigonometric identities or formulas promotes the efficient solving of complex trigonometric problems (Mednykh, 2002).

Following is a list of a few trigonometric identities:

4.2.1. Pythagorean Identities

The Pythagorean theorem, which was first proposed by the renowned mathematician Pythagoras, possesses a broad range of applications (Csiszár & Matúš, 2012). The Pythagorean identities serve as fundamental components for many algebraic modifications of equations and formulas.

The fundamental connection between the sine function and the cosine function is established by the Pythagorean identity:

$sin^2\theta + cos^2\theta = 1$ **Equation 1**

Where;

$sin^2\theta$ means $(sin\theta)^2$ *and* $cos^2\theta$ *means* $(cos\theta)^2$

The relationship described above is a variant of the Pythagorean theorem, which can be derived from the equation $x^2+y^2=1$, representing the unit circle (Zimba, 2009). Several mathematical problems can be solved in terms of the sine and cosine functions.

From Equation 1, we can derive:

$$sin\theta = \pm\sqrt{1 - cos^2\theta}$$

$$cos\theta = \pm\sqrt{1 - sin^2\theta}$$

Where the sign ($\pm$) depends on the quadrant of $\theta.B$y utilizing these identities, it is possible to represent one trigonometric function in terms of any other trigonometric function, as given below.

$$1 + cot^2\theta = csc^2\theta$$

$$1 + tan^2\theta = sec^2\theta$$

$$sec^2\theta + csc^2\theta = sec^2\theta csc^2\theta$$

The use of exponential notation in these identities showcases a unique feature specific to trigonometry. In mathematical terms, $\sin^2 \theta$ represents the square of the sine of angle θ, which can also be written as $(\sin \theta)^2$. This notation simplifies the process of calculating the sine of angle θ and then squaring that value. However, some mathematicians found the use of parentheses around "sin θ" to be unnecessary. As a result, a consensus emerged to adopt a more concise notation. The superscript 2 following the "sin" function indicates the operation of squaring the entire expression, e.g., $\sin^2 \theta$. This modern identity also extends to other trigonometric functions, including cosine squared θ ($\cos^2 \theta$), tangent squared θ ($\tan^2 \theta$), cotangent squared θ ($\cot^2 \theta$), and so forth (Bryan, 1999; Aleshkevich, 2019).

4.2.2. Mother of all Pythagorean Identities

The Pythagorean identity, which gave rise to the other two identities, is expressed as the equation: $\sin^2\theta + \cos^2 \theta = 1$. One may question the origins of this identity and its significance. However, the fundamental Pythagorean identity is the foundation of an equivalence between various trigonometric functions, resulting in a value of 1 (Bernhart & Price, 2007). This simplification proves to be quite advantageous when tackling trigonometric equations. Therefore, it may be said that this particular identity is among the most commonly employed. This identity comes from putting a right triangle inside the unit circle and substituting values and equations to come up with a whole new equation (Bornstein, 2017).

One can discover that in a circle, $\theta = y / r \, and \, cos\theta = x / r$, where (x, y) are the coordinates of a point and r is the radius of the circle. The value of x is also the length of the adjacent side of the triangle (horizontal length), and y is the length of the opposite side (vertical length). In a unit circle, the radius is equal to 1. When you substitute 1 for r in the equation, you find that

$$sin\theta = \frac{y}{r} = y \, and \, cos\theta = \frac{x}{1} = x$$

The Pythagorean theorem states that the sum of the squares of the two shorter sides (legs) of a right triangle is equal to the square of the hypotenuse. In mathematical notation, it can be represented as $a^2 + b^2 = c^2$. When considering a right triangle positioned on the unit circle, where the radius (equivalent to the hypotenuse) is equal to 1, then $x^2 + y^2 = 1^2$, as the unit circle has a radius of 1 (Yeo, 2007). By substituting x with cos θ and y with sin θ, and rearranging the terms, the resulting expression becomes $\sin^2 \theta + \cos^2 \theta = 1$ (Bornstein, 2017; Rakes & Zollman, 2014) (Figure 4.1).

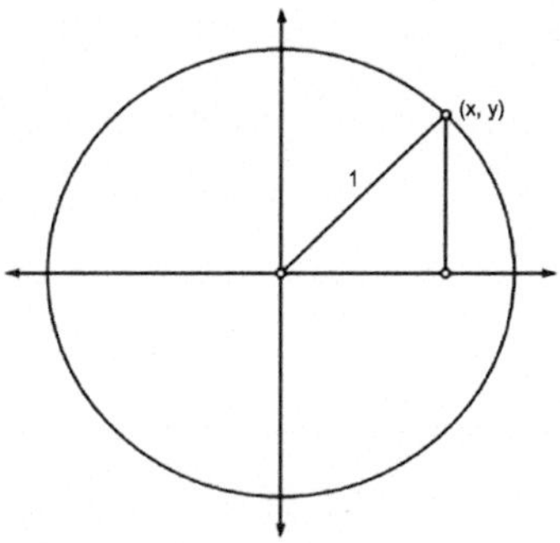

Figure 4.1. Finding the Pythagorean identity on a unit circle.

Source: Jeffrey, Creative Commons License.

Considering an angle of 30 degrees and utilizing the numerical values associated with the trigonometric equations of a 30-degree position, we get:

$$\sin 30° = \frac{1}{2} \, and \, \cos 30° = \sqrt{3} / 2,$$

and putting them into the identity, you get

$$sin^2\theta + cos^2\theta = 1$$

$$(\frac{1}{2})^2 + (\frac{\sqrt{3}}{2})^2 = \frac{1}{4} + \frac{3}{4} = 1$$

4.2.3. Extending to Tangent and Secant

Pythagorean identity (involving tanθ and secθ) is derived from the initial identity that involves sine and cosine. The process involves using algebraic techniques, such as using reciprocal and ratio identities, followed by simplification and introduction of two more identities. Here's the process:

1. Starting with the first Pythagorean identity, $sin^2\theta + cos^2\theta = 1$, divide each term by $cos^2\theta$. $\frac{sin^2\theta}{cos^2\theta} + \frac{cos^2\theta}{cos^2\theta} = \frac{1}{cos^2\theta}$

2. Rewrite each term by using the exponential rule: $\left(\frac{a^2}{b^2}\right) = (a/b)^2$
$(\frac{sin\theta}{cos\theta})^2 + (\frac{cos\theta}{cos\theta})^2 = (1/cos\theta)^2$

3. Replace each of the terms with an equivalent expression.

i. Use a ratio identity, $sin\theta / cos\theta = tan\theta$, to replace the first term with tan θ.

ii. Substitute the second term with 1, i.e., $\frac{cos\theta}{cos\theta} = 1$

iii. Use a reciprocal identity, $1/cos\theta = sec\theta$, to replace the third term with sec θ.

Substituting these expressions into the equation and simplifying, you get

$$(tan\theta)^2 + (1)^2 = (sec\theta)^2$$

$$tan^2 + 1 = sec^2\theta$$

This is the second Pythagorean identity (Loce, 2021).

4.2.4. Finishing up with Cotangent and Cosecant

The other Pythagorean identity (involving cotθ and cscθ) is also derived from the initial identity that involves sine and cosine

The following procedure outlines the steps:

1. Starting with the first Pythagorean identity, $\sin^2\theta + \cos^2\theta = 1$, divide each time by $\sin^2\theta$

$$\frac{sin^2\theta}{sin^2\theta}+\frac{cos^2\theta}{sin^2\theta}=\frac{1}{sin^2\theta}$$

2. Rewrite each term by using the exponential rule : $\left(\frac{a^2}{b^2}\right)=(a/b)^2$

$$(\frac{sin\theta}{sin\theta})^2+(cos\theta/sin\theta)^2=(1/sin\theta)^2$$

3. Substitute each of the terms with linear functions

i. Substitute the first term with 1, i.e., $\frac{sin\theta}{cos\theta}=1$

ii. Use a ratio identity, $\frac{cos\theta}{sin\theta}=\cot\theta$, to substitute the second term by $\cot\theta$.

iii. Use a reciprocal identity, $\frac{1}{sin\theta}=csc\theta$, to substitute the third time with $\csc\theta$.

iv. Substituting these expressions into the equation and simplifying, you find that the result is:

$$(1)^2+(cot\theta)^2=(csc\theta)^2$$

$$1+cot^2\theta=csc^2\theta$$

4.2.5. Ratio Identities

Trigonometry has two fundamental identities known as ratio identities. The ratio identities create ways to write tangent and cotangent using the other two basic functions, sine and cosine (Bornstein, 2017).

The ratio identities refer to a set of mathematical equations that express the relationships between trigonometric functions in terms of ratios of the sides of a right triangle (Rakes & Zollman, 2014).

$$tan\theta=\frac{sin\theta}{cos\theta} and\ cot\theta=cos\theta/sin\theta$$

For sine, cosine, and tangent:

$$sine = \frac{opposite}{hypotenuse}, cosine = \frac{adjacent}{hypotenuse}, and\ tangent = \frac{opposite}{adjacent}$$

You can use these to get:

$$\frac{sine}{cosine} = \frac{\frac{opp}{hyp}}{\frac{adj}{hyp}} = \frac{opp}{\cancel{hyp}} \cdot \frac{\cancel{hyp}}{adj} = \frac{opp}{adj} = tangent$$

Similarly, because cotangent is the reciprocal of tangent:

$$cotangent = \frac{1}{tangent} = \frac{1}{\frac{sine}{cosine}} = \frac{cosine}{sine}$$

4.2.6. Reciprocal Identities

The simplest and most basic trigonometric identities are those involving the reciprocals of the trigonometric functions. It is important to note that the reciprocal of a given number can be obtained by dividing 1 by that particular number. For instance, the reciprocal of 2 is 1/2. Another way to explain reciprocals is to point out that the product of a number and its reciprocal is 1. In the case of 2 and its reciprocal, 1/2, 2x 1/2 = 1. The same principle goes for trigonometric reciprocals.

The reciprocal identifiers are defined as follows:

i. The reciprocal of sine is cosecant: $\frac{1}{sin\theta} = csc\theta$

ii. The reciprocal of cosine is secant: $\frac{1}{\cos\theta} = sec\theta$

iii. The reciprocal of tangent is cotangent: $\frac{1}{tan\theta} = cot\theta$

iv. The reciprocal of cotangent is tangent: $\frac{1}{cot\theta} = tan\theta$

v. The reciprocal of secant is cosine: $\frac{1}{sec\theta} = cos\theta$

vi. The reciprocal of cosecant is sine: $\frac{1}{csc\theta} = sin\theta$

Once you multiply the reciprocal, you get 1:

$$sin\theta \cdot csc\theta = 1$$

$$cos\theta \cdot sec\theta = 1$$

$$tan\theta \cdot cot\theta = 1$$

The reciprocal identity is a highly valuable tool in the process of solving trigonometric equations, particularly those that involve fractional expressions (Kelly, 1988).

4.2.7. Opposite-Angle Identities

The opposite-angle identities are mathematical formulas that transform signs involving negative angles into equivalent expressions involving positive angles. Negative angles are great for describing a situation but aren't handy when sticking them in a trigonometric function and calculating that value. So, for example, you can rewrite the sine of –30 degrees as the sine of 30 degrees by putting a negative sign before the function:

$$\sin(-30°) = -\sin(30)$$

This identity is applicable to angles of varying measures as well. The angle measurement is not limited to -30 degrees; every negative angle is valid.

The opposite-angle identities for the three most basic functions are:

$$\sin(-\theta) = -sin\theta$$

$$\cos(-\theta) = cos\theta$$

$$\tan(-\theta) = -tan\theta$$

The relationship between sine and tangent for negative angles may seem intuitive, but understanding the role of cosine in this context is crucial. How does the cosine of a negative angle relate to its equivalent positive angle? This can be explained by considering the signs of these functions in different quadrants. The sine function is positive when the terminal side of an angle lies between the first and third quadrants, while the cosine function is positive exclusively in the first and fourth quadrants (Sánchez, Gómez, del Rincón, & Musonda, 2023). In the Cartesian coordinate system, positive angles are conventionally measured counterclockwise from the positive side of the x-axis, whereas negative angles are measured clockwise (Mphuthi, 2019). To illustrate this, refer to Figure 4.2, which depicts both a –45-degree angle and a 45-degree angle.

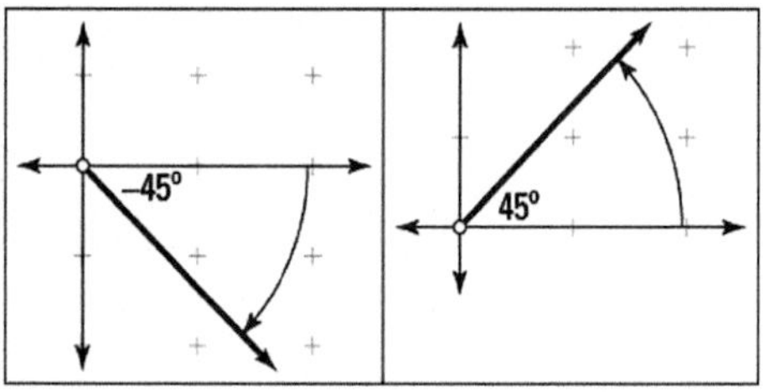

Figure 4.2. Angles of –45 degrees and 45 degrees.

Source: Agarwal, Creative Commons License.

Firstly, consider an angle of –45 degrees in the fourth quadrant, resulting in a negative sine value for its terminal side. Conversely, a 45-degree angle in the first quadrant yields a positive sine value. Now, what happens when we examine a negative angle in the second quadrant, such as -200 degrees? A +200-degree angle yields a negative sine value due to its terminal side residing in the third quadrant. Conversely, a -200-degree angle yields a positive sine value since its terminal side is in the second quadrant. This reveals an interesting relationship: the magnitudes of the sines are inversely related. Thus, it underscores the principle that the sine value of a negative angle equals the negation of the sine value of its corresponding positive angle.

Now, turning our attention to the cosine function, when we analyze its behavior in the coordinate plane, we find that a –45-degree angle exhibits a positive cosine value akin to its counterpart, the 45-degree angle. This parallel behavior underscores the significance of the cosine function in understanding the properties of angles.

$$\cos(-45°) = \cos(45°) = \frac{\sqrt{2}}{2}$$

It is observed that the cosine of an angle that is negative is equivalent to the cosine of the corresponding positive angle.

Subsequently, endeavor to apply the identity from an alternative perspective, specifically a negative angle positioned in the third quadrant with its terminal side. Figure 4.3 depicts an angle with a negative measurement of –120 degrees and its equivalent positive angle of 120 degrees. The angle measuring –120 degrees is situated in the third quadrant, resulting in negative values for both its cosine and sine functions. The angle spanning 120 degrees is characterized by its terminal side located in the second quadrant, wherein the sine function yields a positive value, and the cosine function yields a negative value. The sine of –120 degrees is equal to the negation of the sin of 120 degrees, whereas the cosine of –120 degrees is equivalent to the angle of 120 degrees. In trigonometric notation, the representation appears as follows:

$$\sin(-120°) = -\sin(120°) = -\frac{\sqrt{3}}{2} \textit{ and } \cos(-120°) = \cos(120°) = -\frac{1}{2}$$

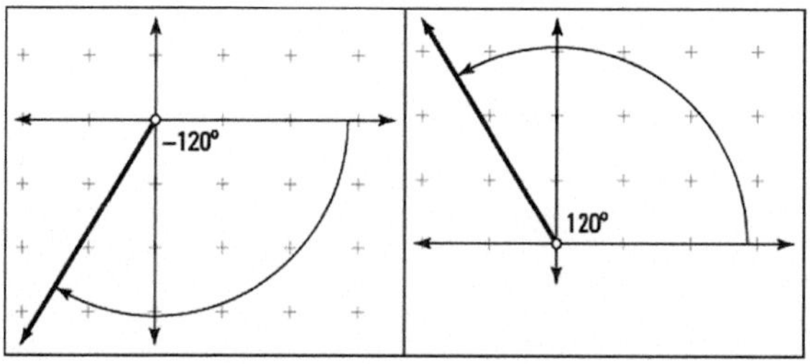

Figure 4.3. Angles of –120 degrees & 120 degrees.

Source: Jeffrey, Creative Commons License.

When the inverse-angle identity is applied to the tangential of a 120-degree angle, it yields a negative value. Consequently, it can be observed that the reverse of a negative value is a positive value. Unexpectedly, the outcome is as anticipated. By utilizing the identity, the application of the opposite operation results in a positive value for the tangent function. This outcome is observed while evaluating the tangent of an angle measuring 120 degrees, whereby the terminal side is situated in the third quadrant, thus yielding a positive value. However, it is possible to examine the inverse-angle identification for the tangent from a different perspective by employing the ratio identity as a means of substantiating its validity.

$$\tan(-\theta) = \frac{\sin(-\theta)}{\cos(-\theta)} = -\frac{\sin(\theta)}{\cos(\theta)} = -\frac{\sin(\theta)}{\cos(\theta)} = -\tan(\theta)$$

Did You Know?

The utilization of double-angle formulae enables the identification of symmetrical patterns observed in nature, like the recurring arrangement of petals in a flower.

4.2.8. Trigonometric Identities of Complementary Angles

In the field of geometry, it is known that two angles are considered complimentary when the sum of their measures is equivalent to 90 degrees (Mickey & McClelland).

Trigonometric identities for complementary angles are given below:

$$Sin\left(90-\theta\right)=Cos\,\theta$$

$$cos\left(90-\theta\right)=Sin\,\theta$$

$$Tan\left(90-\theta\right)=Cot\,\theta$$

$$Cot\left(90-\theta\right)=Tan\,\theta$$

$$Sec\left(90-\theta\right)\quad Csc\,\theta$$

$$Csc\left(90-\theta\right)=Sec\,\theta$$

4.2.9. Trigonometric Identities of Supplementary Angles

Supplementary angles are defined as a pair of angles whose sum equals 90 degrees.

Trigonometric identities for supplementary angles are given below:

$$sin\left(180°-\theta\right)=sin\theta$$

$$cos\left(180°-\theta\right)=-cos\theta$$

$$cosec\left(180°-\theta\right)=cosec\theta$$

$$sec\left(180°-\theta\right)=-sec\theta$$

$$tan(180° - \theta) = -tan\theta$$

$$\cot(180° - \theta) = -\cot\theta$$

4.3. SUM AND DIFFERENCE OF ANGLES OF TRIGONOMETRIC IDENTITIES

Consider two angles, α and β, the trigonometric sum and difference identities are as follows:

$$sin(\alpha + \beta) = sin(\alpha).cos(\beta) + cos(\alpha).sin(\beta)$$

$$sin(\alpha - \beta) = sin\alpha.cos\beta - cos\alpha.sin\beta$$

$$cos(\alpha + \beta) = cos\alpha.cos\beta - sin\alpha.sin\beta$$

$$cos(\alpha - \beta) = cos\alpha.cos\beta + sin\alpha.sin\beta$$

$$\tan(\alpha + \beta) = (tan\alpha + tan\beta) / (1 - \tan\alpha\ .tan\beta)$$

$$\tan(\alpha - \beta) = (tan\alpha - tan\beta) / (1 + \tan\alpha\ .tan\beta)$$

4.3.1. Double Angle Trigonometric Identities

If the angles are doubled, then the trigonometric identities for sin, cos and tan are:

$$sin\,2\theta = 2\,sin\theta\,cos\theta$$

$$cos\,2\theta = Cos^2\theta - Sin^2\theta = 2\,Cos^2\theta - 1 = 1 - 2Sin^2\theta$$

$$tan2\theta = (2tan\theta) / (1 - Tan^2\theta)$$

4.3.2. Half Angle Identities

When angles are divided by two, trigonometric identities are as follows:

$$sin(\theta / 2) = \pm\sqrt{(1 - cos\theta) / 2}$$

$$cos(\theta / 2) = \pm\sqrt{(1 + cos\theta)} / 2$$

$$tan(\theta/2) = \pm\sqrt{(1-cos\theta)/(1+cos\theta)}$$

4.3.3. Sum to Product Trigonometric Identities

The product-sum trigonometric identities change the sum or difference of sines or cosines into a product of sines and cosines.

$$SinA + SinB = 2(Sin(A+B)/2).(Cos(A-B)/2)$$

$$CosA + CosB = 2(Cos(A+B)/2).(Cos(A-B)/2)$$

$$SinA - SinB = 2(Cos(A+B)/2).(Sin(A-B)/2)$$

$$CosA - CosB = -2(Sin(A+B)/2).(Sin(A-B)/2)$$

4.3.4. Product to Sum Trigonometric Identities

These identities are:

$$SinA.SinB = [Cos(A-B) - Cos(A+B)]/2$$

$$SinA.CosB = [Sin(A+B) + Sin(A-B)]/2$$

$$CosA.CosB = [Cos(A+B) + Cos(A-B)]/2$$

$$CosA.SinB = [Sin(A+B) - Sin(A-B)]/2$$

4.4. TRIGONOMETRIC IDENTITIES PROOFS

Similarly, an equation that involves trigonometric ratios of an angle represents a trigonometric identity. The following section covers the fundamental trigonometric identities and their proofs. Consider the right angle ΔABC, which is right-angled at B, as shown in the given figure 4.4.

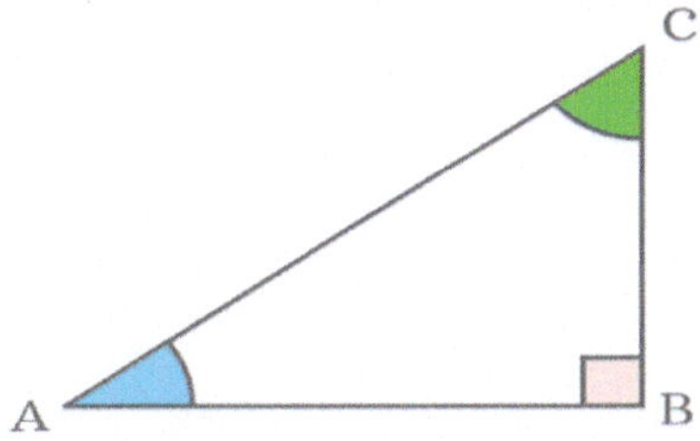

Figure 4.4. Right angled triangle.

Source: Mathdoubts.com, Creative Commons License.

Applying Pythagoras Theorem for the given triangle, we have:

$(\text{hypotenuse})^2 = (\text{base})^2 + (\text{perpendicular})^2$

$AC^2 = AB^2 + BC^2$ (1)

Let us prove the three Pythagoras trigonometric identities, which are commonly used.

4.4.1. Trigonometric Identity 1

Now, divide each term of equation (1) by AC^2, we will have:

$$\frac{AC^2}{AC^2} = \frac{AB^2}{AC^2} + \frac{BC^2}{AC^2}$$

$$\frac{AB^2}{AC^2} + \frac{BC^2}{AC^2} = 1$$

$$(AB / AC)^2 + (BC / AC)^2 = 1(2)$$

We know that,

$$(AB / AC)^2 = \cos a \;\&\; (BC / AC)^2 = \sin a$$

Thus equation (2) can be written as

$$Cos^2\theta + Sin^2\theta = 1$$

Identity 1 is valid for angles $0 \leq a \leq 90$.

4.4.2. Trigonometric Identity 2

Now, dividing the equation (1) by AB^2, we get

$$\frac{AC^2}{AB^2} = \frac{AB^2}{AB^2} + \frac{BC^2}{AB^2}$$

$$\frac{AC^2}{AB^2} = 1 + \frac{BC^2}{AB^2}$$

$$(AC / AB)^2 = 1 + (BC / AB)^2 . (3)$$

By referring to trigonometric ratios, it can be proved that:

$$\frac{AC}{AB} = hypotenuse / side\, adjacent\, to\, angle\, a = \sec a$$

Similarly,

$$\frac{BC}{AB} = side\, opposite\, to\, angle\, a / side\, adjacent\, to\, angle\, a = \tan a$$

Replacing the values of AC/AB and BC/AB in equation (3) gives,

$$1 + tan^2 a = sec^2 a$$

As it is known that tan a is not defined for a = 90°, therefore, identity 2 obtained above is true for $0 \leq A < 90$.

4.4.3. Trigonometric Identity 3

Dividing the equation (1) by BC^2, we acquire

$$\frac{AC^2}{BC^2} = \frac{AB^2}{BC^2} + \frac{BC^2}{BC^2}$$

$$\frac{AC^2}{BC^2} = \frac{AB^2}{BC^2} + 1 (4)$$

By referring to trigonometric ratios, it can be understood that:

$$\frac{AC}{BC} = \frac{Hypotenuse}{side\, opposite\, to\, angle\, a} = cosec\, a$$

Likewise;

$$\frac{AB}{BC} = \frac{side\, adjacent\, to\, angle\, a}{side\, opposite\, to\, angle\, a} = cot\, a$$

Replacing the values of AC/BC and AB/BC in the equation (4) gives,

$$cosec^2 a = 1 + cot^2 a$$

Since cosec a and cot a are not defined for a = 0°. Therefore, the identity 3 obtained is true for all the values of 'a' except at a = 0°. Therefore, the identity is true for all angles such that $0° < a \leq 90°$.

4.5. TRIANGLE IDENTITIES (SINE, COSINE, TANGENT RULE)

If the identities or equations are applicable for all the triangles and not just for right triangles, then they are the triangle identities. These identities will include:

i. Sine law

ii. Cosine law

iii. Tangent law

If A, B and C are the vertices of a triangle and a, b and c are the respective sides, then;

According to the sine law or sine rule;

$$\frac{a}{sinA} = \frac{b}{sinB} = \frac{c}{sinC}$$

Or

$$\frac{sinA}{a} = \frac{sinB}{b} = \frac{sinC}{c}$$

According to cosine law,

$$c^2 = a^2 + b^2 - 2abcos\, C$$

Or

$$cosC = (a^2 + b^2 - c^2) / 2ab$$

According to tangent law,

$$\frac{a-b}{a+b} = \frac{\tan\left(\frac{A-B}{2}\right)}{\tan\left(\frac{A+B}{2}\right)}$$

4.6. SUM AND DIFFERENCE FORMULAS

We have six main sum and difference formulas for the trigonometric functions, including the sine function, cosine function, and tangent function. These formulas help us to evaluate the value of the trigonometric functions at angles, which can be expressed as the sum or difference of special angles 0°, 30°, 45°, 60°, 90°, and 180°. The list of sum and difference formulas is as follows:

sin(A + (b) = sinA cosB + cosA sinB

sin(A – (b) = sinA cosB – cosA sinB

cos(A + (b) = cosA cosB – sinA sinB

cos(A – (b) = cosA cosB + sinA sinB

tan(A + (b) = (tanA + tanB) / (1 – tanA tanB)

tan(A – (b) = (tanA – tanB) / (1 + tanA tanB)

As discussed earlier, we have mainly six sum and difference identities of the sine function, cosine function, and tangent function. Now, we will prove them one by one. Consider a unit circle in a coordinate plane. We know that the coordinates of points on the unit circle are given by (cosθ, sinθ). Now, in the unit circle given below, point P makes an angle α with the positive x-axis and has coordinates (cos α, sin α), whereas point Q makes an angle β with the positive x-axis and has coordinates (cos β, sin β). Observe that the angle POQ is equal to α – β.

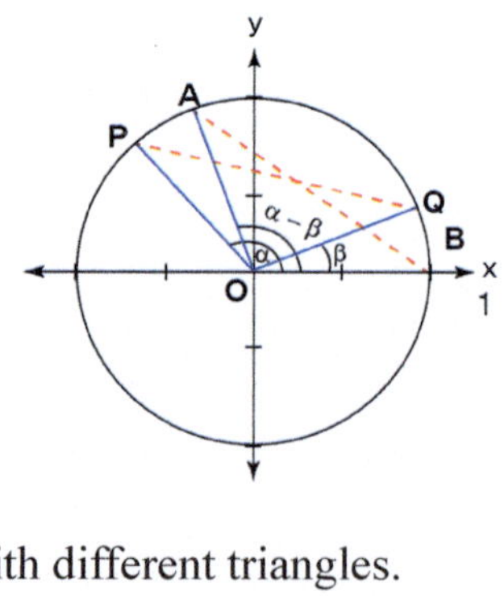

Figure 4.5. Unit circle with different triangles.

Source: Cuemath, Creative Commons License.

Next, label two points A and B on the circle such that point B is on the x-axis with coordinates (1, 0) and angle AOB is equal to α – β and hence, point A has coordinates (cos (α – β), sin (α – β)). Note that triangle POQ and

AOB are rotations of one another, and hence, the lengths of PQ and AB are the same, that is, the distance from P to Q is equal to the distance from A to B (Figure 4.5).

4.6.1. Sum and Difference Formulas for Cosine

First, we will prove the difference formula for the cosine function. Since PQ is equal to AB, using the distance formula, the distance between the points P and Q is given by:

$dPQ = \sqrt{[(\cos \alpha - \cos \beta)2 + (\sin \alpha - \sin \beta)2]}$

$= \sqrt{[\cos 2\alpha - 2 \cos \alpha \cos \beta + \cos 2\beta + \sin 2\alpha - 2 \sin \alpha \sin \beta + \sin 2\beta]}$ --- [Using algebraic identity $(a - b)2 = a2 - 2ab + b2$]

$= \sqrt{[(\cos 2\alpha + \sin 2\alpha) + (\cos 2\beta + \sin 2\beta) - 2 \cos \alpha \cos \beta - 2 \sin \alpha \sin \beta]}$ --- [Using trigonometric formula $\cos 2A + \sin 2A = 1$]

$= \sqrt{[1 + 1 - 2 \cos \alpha \cos \beta - 2 \sin \alpha \sin \beta]}$

$= \sqrt{(2 - 2 \cos \alpha \cos \beta - 2 \sin \alpha \sin \beta)}$ **(1)**

Now, the distance between the points A and B is given by:

$dAB = \sqrt{[(\cos (\alpha - \beta) - 1)2 + (\sin (\alpha - \beta) - 0)2]}$

$= \sqrt{[\cos 2(\alpha - \beta) + 1 - 2 \cos (\alpha - \beta) + \sin 2(\alpha - \beta)]}$

$= \sqrt{(\cos 2(\alpha - \beta) + \sin 2(\alpha - \beta) + 1 - 2 \cos (\alpha - \beta))}$

$= \sqrt{(1 + 1 - 2 \cos (\alpha - \beta))}$

$= \sqrt{[2 - 2 \cos (\alpha - \beta)]}$ - **(2)**

Since the distance between P and Q is equal to the distance between A and B, we have dPQ = dAB. So, from (1) and (2), we have;

$\sqrt{(2 - 2 \cos \alpha \cos \beta - 2 \sin \alpha \sin \beta)} = \sqrt{[2 - 2 \cos (\alpha - \beta)]}$

Squaring both sides of the above equation;

$\Rightarrow 2 - 2 \cos \alpha \cos \beta - 2 \sin \alpha \sin \beta = 2 - 2 \cos (\alpha - \beta)$

$\Rightarrow 2 [1 - \cos \alpha \cos \beta - \sin \alpha \sin \beta] = 2 [1 - \cos (\alpha - \beta)]$

$\Rightarrow 1 - \cos \alpha \cos \beta - \sin \alpha \sin \beta = 1 - \cos (\alpha - \beta)$

$\Rightarrow - \cos \alpha \cos \beta - \sin \alpha \sin \beta = - \cos (\alpha - \beta)$

$\Rightarrow \boldsymbol{\cos (\alpha - \beta) = \cos \alpha \cos \beta + \sin \alpha \sin \beta}$

Hence, we have proved the difference formula of the cosine function.

We can derive the sum formula for the cosine function by substituting $\beta = - (-\beta)$ into the difference formula of cos. So, we have;

cos (α + β) = cos [α – (-β)]

= cos α cos (-β) + sin α sin (-β) --- [Using the formula cos (α – β) = cos α cos β + sin α sin β]

= cos α cos β + sin α (-sin β) --- [Because cos(-x) = cos x and sin(-x) = -sin x]

= cos α cos β – sin α sin β

⇒ **cos (α + β) = cos α cos β – sin α sin β**

Thus, we have derived the sum and difference formulas for the cosine function.

4.6.2. Sum and Difference Formulas for Sine

Next, we will prove the sum and difference for the sine function using the cofunction identities of trigonometry. We know that sin A = cos (π/2 – (a) and cos A = sin (π/2 – A). So, we have;

sin (α + β) = cos [π/2 – (α + β)]

= cos [π/2 – α – β]

= cos [(π/2 – α) – β]

= cos (π/2 – α) cos β + sin (π/2 – α) sin β

= sin α cos β + cos α sin β

⇒ **sin (α + β) = sin α cos β + cos α sin β**

Now, we can write α – β = α + (-β). So, using the sum formula for the sine function, we have;

sin (α – β) = sin [α + (-β)]

= sin α cos (-β) + cos α sin (-β)

= sin α cos β + cos α (-sin β) --- [Because cos (-x) = cos x and sin (-x) = – sin x]

= sin α cos β – cos α sin β

⇒ **sin (α – β) = sin α cos β – cos α sin β**

Hence, we have proved the sum and difference formulas for the sine function.

4.6.3. Sum and Difference Formulas for Tangent

In this section, we will prove the sum and difference identities for the tangent function. We know that the tangent function can be written as the ratio of the sine and cosine, that is, tan A = sin A / cos A. So, we can write tan (α + β) as;

tan (α + β) = sin (α + β) / cos (α + β)

= (sin α cos β + cos α sin β) / (cos α cos β – sin α sin β) --- [Using sin (α + β) = sin α cos β + cos α sin β and cos (α + β) = cos α cos β – sin α sin β]

Dividing the numerator and denominator by cos α cos β, we have

= [(sin α cos β + cos α sin β) / cos α cos β] / [(cos α cos β – sin α sin β) / cos α cos β]

= (sin α / cos α + sin β / cos β) / [1 – (sin α / cos α) × (sin β / cos β)]

= (tan α + tan β) / (1 – tan α tan β)

⇒ tan (α + β) = (tan α + tan β) / (1 – tan α tan β)

Now, using the above formula, we can find the formula for tan (α – β). So, we have

tan (α – β) = tan [α + (-β)]

= [tan α + tan (-β)] / [1 – tan α tan (-β)] --- [Using tan (A + (b) formula]

= [tan α + (-tan β)] / [1 – tan α (-tan β)] --- [Because tan (-x) = – tan x]

= (tan α – tan β) / (1 + tan α tan β)

⇒ tan (α – β) = (tan α – tan β) / (1 + tan α tan β)

Hence, we have proved the sum and difference formulas of the tangent function.

- Example 1: Find the value of tan (2π/3 + π/4).

Solution: To find the value of tan (2π/3 + π/4), we will use the sum formula of the tangent function.

tan (2π/3 + π/4) = (tan2π/3 + tanπ/4) / (1 – tan2π/3 tanπ/4) --- [Using tan(A + (b) = (tanA + tanB) / (1 – tanA tanB)]

= [(-√3) + 1] / [1 – (-√3) (1)]

= (1 – √3) / (1 + √3)

- Example 2: Evaluate the value of sin (5π/4 – π/6)

Solution: Using the difference formula of sine, we have

sin (5π/4 – π/6) = sin5π/4 cos π/6 – cos5π/4 sin π/6 --- [Using sin(A – (b) = sinA cosB – cosA sinB]

= (-1/√2) (√3/2) – (-1/√2) (1/2)

= -√3/2√2 + 1/2√2

$= (1 - \sqrt{3}) / 2\sqrt{2}$

4.7. DOUBLE ANGLE FORMULAS

We will begin this section by deriving the formulas for sin 2A and cos 2A using the formulas for sin (A + (b) and cos (A + B). The formulas we derive for sin 2A and cos 2A are called double-angle formulas. Here is the derivation of the formula for sin 2A.

$$sin2A = sin(A + A) Write 2A as A + A$$

$$= sin A cos A + cos A sin A Sum formula$$

$$= sin A cos A + sin A cos A Commutative property$$

$$= 2 sin A cos A$$

The last line contributes us our first double-angle formula.

$$sin 2A = 2 sin A cos A$$

The first thing to notice about this formula is that it indicates the 2 in sin 2A cannot be factored out and written as a coefficient. That is,

$$sin 2A \neq 2 sin A$$

For example, if

$$A = 30°, sin 2 \cdot 30° = sin 60° = \sqrt{3} / 2, which is not the same as 2 sin 30° = 2\left(\frac{1}{2}\right) = 1.$$

4.8. HALF-ANGLED FORMULAS

In this section, we will derive formulas for sin (A/2) and cos (A/2). These formulas are called half-angle formulas and are derived from the double-angle formulas for cos 2A.

In double angled formulas section, we developed three ways to write the formula for cos 2A, two of which were:

$$cos 2A = 1 - 2 sin^2 A and cos 2A = 2 cos^2 A - I$$

The choice of the letter we use to denote the angles in these formulas is arbitrary, so we can use x instead of A.

$$cos 2x = 1 - 2 sin^2 x and cos 2x = 2 cos^2 x - I$$

Let us exchange sides in the first formula and solve for sin x.

$$1-2\,sin^2\,x=cos\,2x\,Exchange\,sides$$

$$-2\,sin^2\,x=-1+cos\,2x\,Add\,-1\,to\,both\,sides,$$

$$sin^2 x=\frac{1-cos2x}{2}\,Divide\,both\,sides\,by-2$$

$$\sin x=\pm\sqrt{\frac{1-cos2x}{2}}\,take\,the\,sqaure\,root\,of\,both\,sides$$

Because every value of x can be written as ½ of some other number A, we can replace x with A/2. This is equivalent to saying 2x = A.

$$sin\frac{A}{2}=\pm\sqrt{\frac{1-cosA}{2}}$$

The above equation represents the half-angle formula for sine A/2. To derive the half-angle formula for cos A/2, we initiate the process by solving the equation cos 2x = $2\cos^2 x - 1$ for cosx. This equation relates the double angle (2x) to the cosine of x.

$$cos\frac{A}{2}=\pm\sqrt{\frac{1+cosA}{2}}$$

4.9. IDENTITIES AND FORMULAS INVOLVING INVERSE FUNCTIONS

The solution to our first example combines our knowledge of opposite trigonometric functions with our formula for sin (A + B).

Example: Evaluate sin (arcsin 3/5 + arctan 2) without using a calculator.

Solution: If we let $\alpha=\arcsin\frac{3}{5}\,and\;\beta=arctan2$ we can simplify things

$$\sin\left(\arcsin\frac{3}{5}+\arctan 2\right)=\sin\left(\alpha+\beta\right)=sin\alpha\cos\beta+cos\alpha\,sin\beta$$

Drawing and labeling a triangle for α and another for β we have:

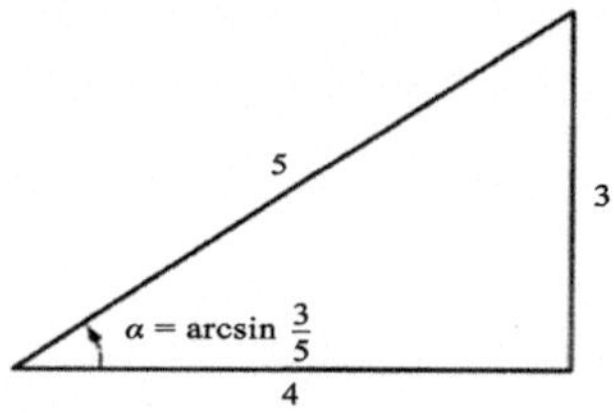

Figure 4.6. Triangle for **á** .

Source: Edward, Creative Commons License.

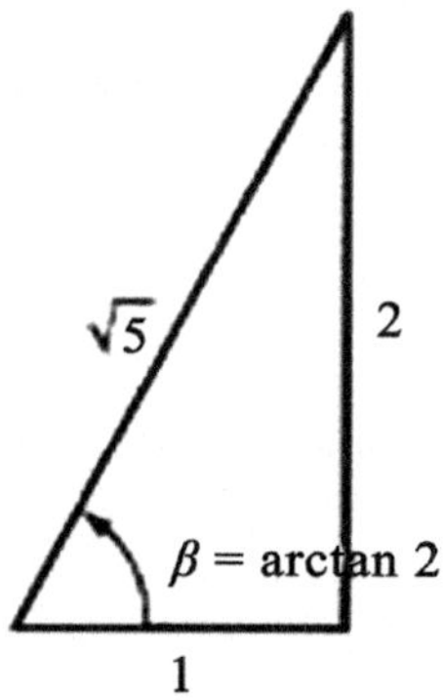

Figure 4.7. Triangle for **â** .

Source: Edward, Creative Commons License.

Based on the triangles shown in Figures 4.6 and 4.7, it can be observed that:

$$sin\alpha = \frac{3}{5} sin\beta = \frac{2}{\sqrt{5}}$$

$$cos\alpha = \frac{4}{5} \cos\beta = \frac{1}{\sqrt{5}}$$

Substituting these numbers into

$sin\alpha \, cos\beta + cos\alpha \, sin\beta$

Result is:

$$\frac{3}{5} \cdot \frac{1}{\sqrt{5}} + \frac{4}{5} \cdot \frac{2}{\sqrt{5}} = \frac{11}{5\sqrt{5}}$$

4.10. TRIGONOMETRIC IDENTITIES AND COMPLEX NUMBERS

The study of trigonometric identities and complex numbers has been a topic of great interest in the field of mathematics. Trigonometric identities are a set of equations that are characterized by the presence of trigonometric functions such as sine, cosine, and tangent. These identities are utilized to simplify expressions and solve equations that involve these functions. Complex numbers, on the other hand, are numbers that exhibit both real and imaginary components. They are typically represented in the form $a + bi$, where a represents the real component, and b represents the imaginary component. The geometric representation of complex numbers in the complex plane is an intriguing aspect of this field. Furthermore, the manipulation of complex numbers utilizing trigonometric functions and identities is an area of active research.

Constants: $e\left(Euler's\ Number\right), i(the\ imaginary\ unit, i^2 = -1$ and

$\pi(pi, the\ ratio\ of\ circle's\ circumference\ to\ its\ diameter\ )$. Euler's formula stated as:

$$e^{i\theta} = \cos\left(\theta\right) + isin\left(\theta\right)$$

- **De Moivre's Formula:** De Moivre's formula, an outcome derived from Euler's formula, establishes a connection between complex exponentials and trigonometric ratios raised to an arbitrary integer power n (Özdemir, 2019). It is represented as follows:

$$e^{ino} = (\cos\left(\theta\right) + isin\left(\theta\right))^n$$

De Moivre's formula empowers us to efficiently compute high powers of trigonometric functions, making complex calculations more manageable. It finds wide applications in areas such as electrical engineering, where it simplifies the analysis of alternating current (AC) circuits (Kösal & Bilgili, 2020).

4.11. APPLICATIONS OF TRIGONOMETRIC IDENTITIES AND FORMULAS

Trigonometric identities and formulae, which stem from the relationship between angles and ratios, have a wide range of applications in many fields (Touk et al., 2017). They contribute to the growth of our comprehension of the natural world and facilitate technological progress (Rathour, Obradovic, Tiwari, Mishra, & Mishra, 2023). As the extensive uses of these mathematical tools are examined, it becomes evident that trigonometry permeates several domains, such as science, navigating, physics, and other related fields (Berndt, 1975).

4.11.1. Engineering and Architecture

Trigonometric identities and formulas are essential in the fields of engineering and architecture. Engineers and architects use these identities to solve complex equations that involve angles and distances. For example, trigonometric identities are used to calculate the dimensions of structures such as buildings and bridges (Maqsood et al., 2021). They are also used in surveying to determine the angles and distances between points on the ground. In addition, architects use trigonometric identities to design buildings with specific angles and shapes. The application of these identities and formulas in engineering and architecture is a testament to their usefulness and importance in these fields (Blanquer et al., 2017).

4.11.2. Navigation and Astronomy

Throughout the course of history, trigonometry has played a key role in the field of navigation, facilitating the ability of seafarers and sailors to ascertain their precise locations while traversing the vast expanse of the sea (Peirce, 1845). Celestial navigation relies on the utilization of trigonometric functions to determine latitude and longitude by means of detecting the angles formed by celestial bodies with the horizon (Farrar, 1833). In addition, trigonometric identities play a crucial role in enabling celestial computations and forecasting astronomical phenomena, hence achieving substantial advancements in the realm of astronomy (Dougharty).

4.11.3. Music and Sound Analysis

Trigonometric identities play a crucial role in the domain of sound and music (Serra & Smith, 1990). The field of harmonic analysis focuses on the examination of intricate sound waves, aiming to decompose them into their constituent frequencies or harmonics (Kromhout, 2020). Trigonometric functions play a crucial role in this analytical process (Mansur, Blattner, & Joy, 1985). Through a comprehensive comprehension of the interconnections among harmonics, individuals in the field of music, including musicians and sound engineers, possess the ability to skillfully control sound waves in order to generate intricate and aesthetically beautiful musical tones (Kronland-Martinet, 1988).

4.11.4. Aviation and Aeronautics

Trigonometry assumes a crucial role within the field of aviation and aeronautics, providing essential information to pilots in the areas of flight navigation, air traffic control, and aircraft design. Trigonometric functions are employed by pilots to ascertain altitude, direction, and angles of climb or fall (Zhang, Peng, Meng, Zhao, & Li, 2023). Furthermore, the field of aircraft design heavily depends on the use of trigonometry in order to enhance wing designs, hence guaranteeing optimal aerodynamic efficiency and stability during the duration of flight (Cinefra et al., 2020; Koglbauer & Braunstingl, 2021).

4.12. SUMMARY

The chapter titled "Identities & Formulas" offers a compelling exploration of the field of trigonometry, delving into the intricate connections between angles and ratios. This summary provides an examination of the main subjects addressed in the chapter, revealing distinct perspectives and tangible implementations of these mathematical concepts.

Trigonometric identities play a fundamental role in the field of trigonometry since they demonstrate significant equalities that arise from the interconnections among trigonometric functions. The Pythagorean identity, regarded as a notable mathematical result, demonstrates a coherent relationship among the cosine and sine functions. The chapter delves into the reciprocal, quotient, and co-function identities, elucidating the complimentary connections that exist between trigonometric functions and their reciprocal values. The topic also explores the sum and difference formulae, which offer effective techniques for representing the trigonometric functions of the total or difference of degrees in terms of separate functions. This facilitates streamlined computations.

Trigonometric identity proofs provide a comprehensive exploration of the sophisticated practice of establishing these identities. The chapter elucidates a range of methodologies employed to ascertain the legitimacy of trigonometric equations, fostering a comprehensive comprehension of the complexities inherent in their derivations.

The utilization of Triangle Identities expands the scope of trigonometry beyond the limitations of right-angled triangles, enabling its application to many sorts of triangles. The aforementioned identities provide a connection between the trigonometric functions of the angles present in a triangle and the lengths of its sides, revealing the significant correlations that exist between angles and ratios in every given triangle.

The investigation of Sum and Difference Formulas provides access to potent mathematical representations for trigonometric functions involving the summation or subtraction of two angles. These formulae have a wide range of applications in the simplification of expressions, the transformation of functions, and the identification of symmetries and periodic patterns.

The utilization of Double Angle Formulas extends our understanding of trigonometry by providing a means to define trigonometric functions with double angles in relation to the original functions. This phenomenon gives rise to thought-provoking observations on symmetrical connections and periodic actions, hence creating novel opportunities for comprehending intricate waveforms and oscillatory events.

The use of Half Angle Formulas provides a refined method for representing trigonometric functions involving half angles, facilitating the simplification of intricate formulas, and revealing symmetrical connections among angles and ratios.

The concept of Identity and Formulas with Inverse Functions serves to construct meaningful relationships between trigonometric equations and their respective inverse functions. These connections are of great value in the process of solving trigonometry equations and simplifying trigonometric expressions.

REFERENCES

1. Abdullah, K. A., Marjudi, S., Yusof, Z., & Sulaiman, R., (2019). *Algebraic-Trigonometric Nonlinear Analytical Inverse Kinematic Modeling and Simulation for Robotic Manipulator Arm Motion Control.* Paper presented at the Advances in Visual Informatics: 6th International Visual Informatics Conference, IVIC 2019, Bangi, Malaysia, Proceedings 6.
2. Abdullah, K. A., Yusof, Z., Ahmad, R. M. T. R. L., Rauf, M. F. A., Adnan, Z., Hassan, W. A. W., & Sulaiman, R., (2022). Algebraic models based on trigonometric and Cramer's rules for computing inverse kinematics of robotic arm. *International Journal of Mechatronics and Automation, 9*(1), 1–11.
3. Adeyemo, O. D., & Khalique, C. M., (2023). Shock waves, periodic, topological kink, and singular soliton solutions of a new generalized two-dimensional nonlinear wave equation of engineering physics with applications in signal processing, electromagnetism, and complex media. *Alexandria Engineering Journal, 73*, 751–769.
4. Aleshkevich, N., (2019). *Geometric and Algebraic Interpretation of Primitive Pythagorean Triples Parameters.* arXiv preprint arXiv:1907.05540.
5. Alquran, M., & Krishnan, E. V., (2016). Applications of sine-Gordon expansion method for a reliable treatment of some nonlinear wave equations. *Nonlinear Stud, 23*(4), 639–649.
6. Berndt, B. C., (1975). Character analogues of the Poisson and Euler-MacLaurin summation formulas with applications. *Journal of Number Theory, 7*(4), 413–445.
7. Bernhart, F., & Price, H. L., (2007). *Heron's Formula, Descartes Circles, and Pythagorean Triangles*. arXiv preprint math/0701624.
8. Blanquer, S. B., Werner, M., Hannula, M., Sharifi, S., Lajoinie, G. P., Eglin, D., & Grijpma, D. W., (2017). Surface curvature in triply-periodic minimal surface architectures as a distinct design parameter in preparing advanced tissue engineering scaffolds. *Biofabrication, 9*(2), 025001.
9. Bornstein, N. M., (2017). *Music of the Triangles: How Students Come to Understand Trigonometric Identities and Transformations.* University of New Hampshire.
10. Bryan, T. J., (1999). The conceptual knowledge of preservice secondary

mathematics teachers: How well do they know the subject matter they will teach. *Issues in the Undergraduate Mathematics Preparation of School Teachers: The Journal, 1,* 1–12.

11. Mona., (2015). Introducing the fourth Pythagorean identity. *Mathematics Teacher, 108*(7), 484-486.
12. Cieśliński, J. L., (2012). New definitions of exponential, hyperbolic and trigonometric functions on time scales. *Journal of Mathematical Analysis and Applications, 388*(1), 8–22.
13. Cinefra, M., D'Amico, G., De Miguel, A., Filippi, M., Pagani, A., & Carrera, E., (2020). Efficient numerical evaluation of transmission loss in homogenized acoustic metamaterials for aeronautical application. *Applied Acoustics, 164*, 107253.
14. Csiszár, I., & Matúš, F., (2012). *Generalized Minimizers of Convex Integral Functionals, Bregman Distance, Pythagorean Identities.* arXiv preprint arXiv:1202.0666.
15. Dan, H., & Fen, W., (2019). Application of precision trigonometric elevation measurement in river-crossing leveling based on autocollimation total station. *Bulletin of Surveying and Mapping*, (3), 159.
16. De Gersem, H., Ion, M., Wilke, M., Weiland, T., & Demenko, A., (2006). Trigonometric interpolation at sliding surfaces and in moving bands of electrical machine models. *COMPEL-the International Journal for Computation and Mathematics in Electrical and Electronic Engineering, 25*(1), 31–42.
17. Dougharty, J., (1747) *Mathematical Digests: Containing the Elements and Application of Geometry, and Plane Trigonometry: Whether by Instrumental Construction, or by Calculation, to the Measuring of Heights and Distances, &c.: And the Stereographic Projection of Spheric Trigonometry; With Numerical Solutions, and the Application Thereof, to Several Curious and Important Problems in Astronomy, Navigation, and Dialing: With Tables for Finding the Place, and Eclipses of the Sun and Moon, According to the Last Improvement of the Newtonian Theory; and Many Practical Problems in Each Branch,* 1, 10-19.
18. Elhadad, M., El-Dolil, S. A., & Albagory, Y. A., (2009). *Application of Trigonometric Transforms in Discrete Multi-Tone Systems.* Paper presented at the 2009 International Conference on Computer

Engineering & Systems.

19. Erukala, M. D. C. R. B., & Kumar, K. S., (2021). Multi-level thresholding for image segmentation on medical images using multi-Otsu and sine cosine optimization algorithm. *Annals of the Romanian Society for Cell Biology*, 12305–12316.
20. Farrar, J., (1833). *An Elementary Treatise on the Application of Trigonomentry to Orthographic and Stereographic Projection, Dialing, Mensuration of Heights and Distances, Navigation, Nautical Astronomy, Surveying and Levelling: Together with Logarithmic and Other Tables: Designed for the Use of the Students of the University at Cambridge, New England*. Hilliard, Gray and Company.
21. Gervois, A., & Mehta, M., (1995). Some trigonometric identities encountered by McCoy and Orrick. *Journal of Mathematical Physics, 36*(9), 5098–5109.
22. Gur, H., (2009). Trigonometry learning. *New Horizons in Education, 57*(1), 67–80.
23. Hast, A., Barrera, T., & Bengtsson, E., (2003). *Shading by Spherical Linear Interpolation Using de Moivre's Formula, 3(7), 57-60.*
24. Heng, Z., Bo, H., & Changzheng, Y., (2019). Research and application of precise trigonometric leveling to replace second-class leveling achieving river-crossing leveling. *Bulletin of Surveying and Mapping,* (11), 121.
25. Iqbal, M., Seadawy, A. R., & Lu, D., (2019). Applications of nonlinear longitudinal wave equation in a magneto-electro-elastic circular rod and new solitary wave solutions. *Modern Physics Letters B, 33*(18), 1950210.
26. Jakubský, V., Znojil, M., Luís, E. A., & Kleefeld, F., (2005). Trigonometric identities, angular Schrödinger equations and a new family of solvable models. *Physics Letters A, 334*(2, 3), 154–159.
27. John, S., & Kumar, S., (2023). IoT based medical image encryption using linear feedback shift register–Towards ensuring security for teleradiology applications. *Measurement: Sensors, 25*, 100676.
28. Kelly, A. E., (1988). *The Effects of Search and Heuristics on the Learning of Proofs of Trigonometric Identities.* Stanford University.
29. Khatun, M. A., Arefin, M. A., Uddin, M. H., Baleanu, D., Akbar, M. A., & İnç, M., (2021). Explicit wave phenomena to the couple type

fractional order nonlinear evolution equations. *Results in Physics, 28,* 104597.

30. Khuzwayo, N. Z., (2019). *Exploring Grade 11 Mathematics Learners Learning of Trigonometric Identities: A Case Study of One School in KwaZulu-Natal, 1, 1-19.*
31. Kilar, N., & Simsek, Y., (2021). Computational formulas and identities for new classes of Hermite-based Milne–Thomson type polynomials: Analysis of generating functions with Euler's formula. *Mathematical Methods in the Applied Sciences, 44*(8), 6731–6762.
32. Kim, J., (2017). A representation of de Moivre's formula over Pauli quaternions. *Ann. Acad. Rom. Sci. Ser. Math. Appl., 9*(2), 145–151.
33. Kiradjiev, K. B., (2020). 104.31 polygons and complex trigonometric identities without complex numbers. *The Mathematical Gazette, 104*(561), 522–527.
34. Koglbauer, I., & Braunstingl, R., (2021). Applications of cardiac and electrodermal activity assessment in aviation. In: Koglbauer, I., & Biede-Straussberger, S., (eds.), *Aviation Psychology: Applied Methods and Techniques* (pp. 141–162).
35. Kösal, H. H., & Bilgili, T., (2020). Euler and De Moivre's Formulas for fundamental matrices of commutative quaternions. *International Electronic Journal of Geometry, 13*(2), 98–107.
36. Kromhout, M. J., (2020). The unmusical ear: Georg Simon ohm and the mathematical analysis of sound. *ISIS, 111*(3), 471–492.
37. Kronland-Martinet, R., (1988). The wavelet transform for analysis, synthesis, and processing of speech and music sounds. *Computer Music Journal, 12*(4), 11–20.
38. Loce, M., (2021). *Using Dynamic Graphing Software to Develop Conceptual Understanding of Transformations of Trigonometric Functions*. SUNY Brockport, Department of Education and Human Development,
39. Mall, K., Nolan, S. M., Levin, W. C., Risany, L., & DeLaurentis, D. A., (2023). *Using Uniform Trigonometrization Method for Aviation Based Optimal Control Problems.* Paper presented at the AIAA Aviation 2023 Forum.
40. Mansur, D. L., Blattner, M. M., & Joy, K. I., (1985). Sound graphs: A numerical data analysis method for the blind. *Journal of Medical Systems, 9*, 163–174.

41. Maqsood, S., Abbas, M., Miura, K. T., Majeed, A., & Iqbal, A., (2020). Geometric modeling and applications of generalized blended trigonometric Bézier curves with shape parameters. *Advances in Difference Equations, 2020*, 1–18.

42. Maqsood, S., Abbas, M., Miura, K. T., Majeed, A., Hu, G., & Nazir, T., (2021). Shape-adjustable developable generalized blended trigonometric Bézier surfaces and their applications. *Advances in Difference Equations, 2021*(1), 1–32.

43. Mednykh, A. D., (2002). *Trigonometric Identities and Geometrical Inequalities for Links and Knots.* Paper presented at the Proceedings of the Third Asian Mathematical Conference 2000.

44. Mickey, K., & McClelland, J. L., (2013) *Running Circles Around Symbol Manipulation: The Role of Visualization in Understanding Trigonometric Identities, 1, 1-28.*

45. Mphuthi, G. T., (2019). *Exploring Teachers' Mathematical Meanings for Teaching Trigonometric Ratios and Functions to Grade 10 Learners in the Tshwane South District of Gauteng Province, 1, 1-20.*

46. Nabil, M., Mahfouz, D. M., & Shehata, O. M., (2022). *Development and Evaluation of a Control Architecture for Human-Collaborative Robotic Manipulator in Industrial Application.* Paper presented at the 2022 14th International Conference on Computer and Automation Engineering (ICCAE).

47. Ni, L., Jung, H. Y., & Zhou, Z., (2015). *Enhancing First-year Engineering Students' Trigonometry Learning Experience.* Paper presented at the 2015 ASEE Annual Conference & Exposition.

48. Ningrum, F. V., Nugroho, P. B., & Wulandari, I., (2015) *The Integration of History, Technology, Engineering, Art, and Science into Trigonometry of Senior High/Vocational Middle School, 1, 1-6.*

49. Özdemir, M., (2019). Finding n-th roots of a 2× 2 real matrix using De Moivre's formula. *Advances in Applied Clifford Algebras, 29*(1), 2.

50. Pascual, L. E., Gamit, A. M., Mina, J. C., Campos, Jr. R. B., & Subia, G. S., (2020). Engineering students' retention and transfer of learning from trigonometry to electric circuits. *PalArch's Journal of Archaeology of Egypt/Egyptology, 17*(7), 9609–9613.

51. Peirce, B., (1845). *An Elementary Treatise on Plane & Spherical Trigonometry: With Their Applications to Navigation, Surveying, Heights, and Distances, and Spherical Astronomy, and Particularly*

Adapted to Explaining the Construction of Bowditch's Navigator, and the Nautical Almanac. J. Munroe, 1, 1-17.

52. Radwan, A. G., & Elwakil, A. S., (2012). An expression for the voltage response of a current-excited fractance device based on fractional-order trigonometric identities. *International Journal of Circuit Theory and Applications, 40*(5), 533–538.
53. Rakes, M., & Zollman, A., (2014). The honors dichotomy: Characterizing students in a US high school honors precalculus class. *The Mathematics Educator, 15*(2), 31–49.
54. Rathour, L., Obradovic, D., Tiwari, S., Mishra, L., & Mishra, V., (2023). Trigonometric functions and some trigonometrical identities. *Computational Algorithms and Numerical Dimensions, 2*(1), 1–6.
55. Ratnaningsih, N., Santika, S., Hidayat, E., & Nuraeni, R., (2022). *Development of Interactive Multimedia Using Adobe Flash Based on Problem Solving in Trigonometry Learning*. Paper presented at the AIP Conference Proceedings.
56. Rohimah, S. M., & Prabawanto, S., (2019). Student's difficulty identification in completing the problem of equation and trigonometry identities. *International Journal of Trends in Mathematics Education Research, 2*(1), 34–36.
57. Ryan, C., O'Neill, M., & Collins, J., (1998). *Grammatical Evolution: Solving Trigonometric Identities.* Paper presented at the proceedings of Mendel.
58. Sánchez, M. A., Gómez, C. P., Del Rincón, T. O., & Musonda, T. M., (2023). Designing a theoretical proposal using problem–based learning to improve learning of trigonometric ratios among grade eleven students. *American Journal of Educational Research, 11*(2), 53–78.
59. Seadawy, A. R., & Jun, W., (2017). Mathematical methods and solitary wave solutions of three-dimensional Zakharov-Kuznetsov-burgers equation in dusty plasma and its applications. *Results in Physics, 7*, 4269–4277.
60. Serra, X., & Smith, J., (1990). Spectral modeling synthesis: A sound analysis/synthesis system based on a deterministic plus stochastic decomposition. *Computer Music Journal, 14*(4), 12–24.
61. Ssebaggala, L., (2019). *Investigating Pre-Service Secondary Mathematics Teachers' Reasoning When Learning Trigonometry Using the Line Segment Approach*. Illinois State University.

62. Touk, S. A., Al Houchan, Z., & El Bachraoui, M., (2017). Proofs for two q-trigonometric identities of gosper. *Journal of Mathematical Analysis and Applications, 456*(1), 662–670.
63. Tuan, H. D., Son, T. T., Vo, B., & Nguyen, T. Q., (2005). *LMI Characterization for the Convex Hull of Trigonometric Curves and Applications.* Paper presented at the Proceedings. (ICASSP'05). IEEE International Conference on Acoustics, Speech, and Signal Processing, 2005.
64. Wahyuningsih, S., (2022). Developing android-based interactive-module on trigonometry to enhance the learning motivation of students. *International Journal of Interactive Mobile Technologies, 16*(02), 159.
65. Weber, K., (2008). Connecting research to teaching: Teaching trigonometric functions: Lessons learned from research. *The Mathematics Teacher, 102*(2), 144–150.
66. Yahya, N. A., Varatharajoo, R., & Harithuddin, A., (2020). Satellite formation flying relative geodesic and latitudinal error measures. *Journal of Aeronautics, Astronautics and Aviation, 52*(1), 83–93.
67. Yeo, A. N. H., (2007). *Trig Or Treat: An Encyclopedia of Trigonometric Identity Proofs (TIPs) with Intellectually Challenging Games*. World Scientific.
68. Yokuş, A., Durur, H., Nofal, T. A., Abu-Zinadah, H., Tuz, M., & Ahmad, H., (2020). Study on the applications of two analytical methods for the construction of traveling wave solutions of the modified equal width equation. *Open Physics, 18*(1), 1003–1010.
69. Zhang, Y., Peng, D., Meng, G., Zhao, Q., & Li, T., (2023). *Simulation of Fault Diagnosis Model for Managing Aeronautical Multivariate Heterogeneous Inputs.* Paper presented at the 2023 International Conference on Pattern Recognition, Machine Vision, and Intelligent Algorithms (PRMVIA).
70. Zimba, J., (2009). *On the Possibility of Trigonometric Proofs of the Pythagorean Theorem*. Paper Presented at the Forum geometricorum.

CHAPTER 5

Triangles

Contents

5.1. INTRODUCTION

A triangle is a polygon with three edges and three vertices. It is one of the basic shapes in geometry. A triangle with vertices *A*, *B*, and *C* is denoted ΔABC. In Euclidean geometry, any three points, when non-collinear, determine a unique triangle aand, simultaneously, a unique plane (i.e., a two-dimensional Euclidean space). In other words, there is only one plane that contains that triangle, and every triangle is contained in some plane. If the entire geometry is only the Euclidean plane, there is only one plane, and all triangles are contained in it; however, in higher-dimensional Euclidean spaces, this is not correct (Yiu, 2001). The terminology for categorizing triangles is more than two thousand years old, having been defined on the very first page of Euclid's Elements. The names used for modern classification are either a direct transliteration of Euclid's Greek or their Latin translations (Nipper & Sztajn, 2008).

The right-angled triangle is a specific example of a triangle that offers geometrical insights, making it one of the most fascinating features of triangles (Weber et al., 2012). Using the well-known Pythagorean Theorem, we will demonstrate the relationship between the lengths of a right-angled triangle's sides. We will come across real-world examples in this chapter that show how important triangles are to everyday life. Navigators use celestial triangles to navigate their ships, sturdy buildings are built by architects, and figure out the forces operating on bridges thanks to the potent concepts of triangles and trigonometry (Darvasi, 2003).

5.2. TYPES OF TRIANGLES

There are different types of triangles in math that can be distinguished based on their sides and angles (Figure 5.1).

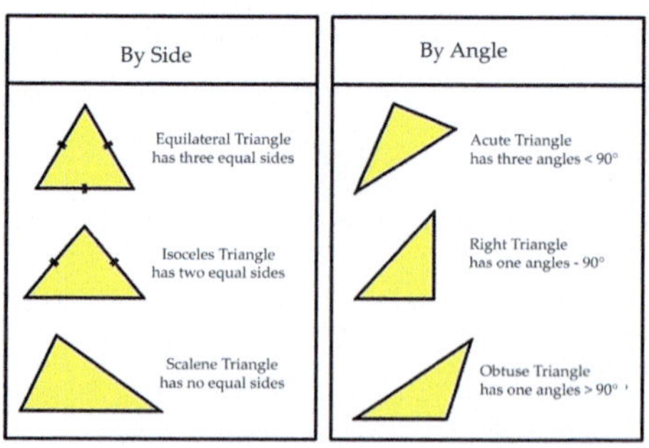

Figure 5.1. Types of triangles based on sides and angles.

Source: Cuemath, Creative Commons.

The characteristics of a triangle's sides and angles are used to classify them. The different types of triangles are as follows:

5.2.1. Types of Triangles Based on Sides

On the basis of side lengths, the triangles are classified into the following types:

- **Equilateral Triangle:** A triangle is considered to be an equilateral triangle when all three sides have the same length.
- **Isosceles triangle:** When two sides of a triangle are equal or congruent, then it is called an isosceles triangle.
- **Scalene triangle:** When none of the sides of a triangle are equal, it is called a scalene triangle.

5.2.2. Types of Triangles Based on Angles

On the basis of angles, triangles are classified into the following types:

- **Acute Triangle:** When all the angles of a triangle are acute, that is, they measure less than 90°, it is called an acute-angled triangle or acute triangle.
- **Right Triangle:** When one of the angles of a triangle is 90°, it is called a right-angled triangle or right triangle.
- **Obtuse Triangle:** When one of the angles of a triangle is an obtuse angle, that is, it measures greater than 90°, it is called an obtuse-angled triangle or obtuse triangle.

5.2.3. Types of Triangles Based on Sides and Angles

The different types of triangles are also classified according to their sides and angles as follows:

- **Obtuse Isosceles Triangle:** A triangle in which 2 sides are equal and one angle is an obtuse angle is called an obtuse isosceles triangle.
- **Isosceles Right Triangle:** A triangle in which 2 sides are equal and one angle is 90° is called an isosceles right triangle. So, in an isosceles right triangle, two sides and two acute angles are congruent.

- **Right Scalene Triangle:** A triangle in which any one of the angles is a right angle and all the 3 sides are unequal is called a right scalene triangle.
- **Acute Isosceles Triangle:** A triangle in which all 3 angles are acute angles and 2 sides measure the same is called an acute isosceles triangle.
- **Acute Scalene Triangle:** A triangle that has 3 unequal sides and 3 acute angles is called an acute scalene triangle.
- **Obtuse Scalene Triangle:** A triangle with an obtuse angle with sides of different measures is called an obtuse scalene triangle.

5.3. RIGHT ANGLED TRIANGLE

An orthogonal triangle, also known as a right triangle or right-angled triangle, is a triangle with one angle that is right-angled (90-degree), meaning that two of its sides are perpendicular (Stillwell & Stillwell, 1989).

The side opposite to the right angle is called the hypotenuse (side *c* in the figure 5.2). The sides adjacent to the right angle are called legs (Kimberling, 1994). Side *a* may be identified as the side adjacent to angle *B* and opposed to (or opposite) angle *A*, while side *b* is the side adjacent to angle *A* and opposed to angle *B* (Clarke, 2005).

If the lengths of all three sides of a right triangle are integers, the triangle is said to be a **Pythagorean triangle** and its side lengths are collectively known as a Pythagorean triple (Moutsios et al., 2014).

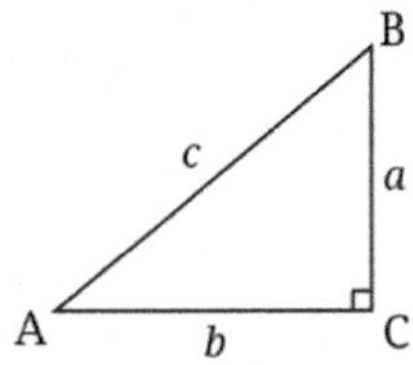

Figure 5.2. Right angled triangle.

Source: Howard, Creative Commons License.

5.3.1. Principal Properties

5.3.1.1. Sides

The Pythagorean theorem establishes a relationship between the three sides of a right triangle (Wise, 2012). In modern algebraic notation, it may be expressed as follows:

$$a^2 + b^2 = c^2$$

In the context of a right triangle, the hypotenuse (the side opposite that of the right angle) is denoted as "c," while the remaining two sides are represented by "a" and "b." (Okumura, 1990). Pythagorean triples are sets of integer values (a, b, c). The theorem under consideration was established during ancient times (Bóna & Tóth, 1996). In right-angled triangles, the square on the side subtending the right angle is equal to the sum of the squares of the sides containing the right angle (Bayu et al., 2021).

5.3.1.2. Area

Like any other triangle, the area may be determined by multiplying half of the base by the equivalent height (Ann, 1988) (Figure 5.3).

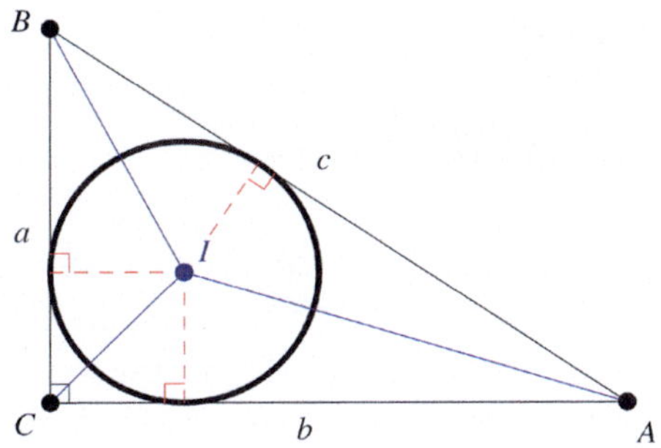

Figure 5.3. Incircle in a right-triangle.

Source: Quora, creative commons.

In the context of a right triangle, it can be seen that when one of the legs is considered as the base, the other leg may be considered as the height (Herbst & Chazan, 2012). Consequently, the area of a triangle with right angles can be determined by multiplying the lengths of the two legs and dividing the product by two (Moldavanov, 2017; Rababa, 2022).

$$A = \frac{1}{2}ab$$

where b represents the base of the triangle, and a indicates the height.

This formula only applies to right triangles (Wise, 2012).

5.3.1.3. Altitudes

The formula for the altitude of a triangle can be derived from the basic formula for the area of a triangle, which is: Area = 1/2 × base × height, where the height represents the altitude. Using this formula, we can derive the altitude formula, which will be, Altitude of the triangle = (2 × Area)/ base.

You can find out the altitude of a scalene triangle, equilateral triangle, right triangle, and isosceles triangle using the following relations (Table 5.1).

Table 5.1. Formulas for Calculating Altitudes of Different Triangles

Scalene Triangle	$h = \frac{2\sqrt{s(s-a)(s-b)(s-c)}}{b}$
Isosceles Triangle	$h = \sqrt{a^2 - \frac{b^2}{4}}$
Equilateral Triangle	$h = \frac{a\sqrt{3}}{2}$
Right Triangle	$h = \sqrt{xy}$

When an altitude is extended from the vertex containing the right angles to the hypotenuse of a triangle, the triangle is divided into two smaller triangles (Goodchild & Sriraman, 2012; Ramachandran, 2016). These smaller triangles exhibit similarity to the original triangle, and thus, they are also similar to each other ((Nader, 2021; **Şengun** & Yilmaz, 2021; Rababa, 2022).

5.3.1.4. Inradius and Circumradius

The radius of the incircle of a right triangle with legs *a* and *b* and hypotenuse *c* is:

$$r = (a+b-c)/2 = ab/(a+b+c)$$

The radius of the circumcircle is half the length of the hypotenuse,

$$R = c/2$$

Thus, the sum of the circumradius and the inradius is half the sum of the legs,

$$R + r = (a + b)/2$$

One of the legs can be expressed in terms of the inradius and the other leg (Zhang et al., 2020).

$$a = 2r(b - r)/(b - 2r)$$

5.3.2. Trigonometric Ratios

The trigonometric functions for acute angles can be defined as ratios of the sides of a right triangle (Yi et al., 2013). A right-angled triangle may be formed with a specific angle, and its sides can be designated as opposed, adjacent, and hypotenuse with respect to this angle (Jumaniyozova, 2020). These ratios of the sides do not depend on the particular right triangle chosen but only on the given angle since all triangles constructed this way are similar. If an angle α is provided, and the opposing side, adjacent side, and hypotenuse are denoted as O, A, and H correspondingly, then trigonometry functions may be defined (Khan, 2020).

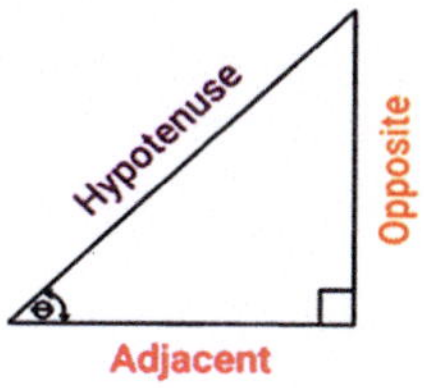

$$sin\alpha = \frac{O}{H}, \cos\alpha = \frac{A}{H}, \tan\alpha = \frac{O}{A}, \sec\alpha = \frac{H}{A}, \cot\alpha = \frac{A}{O}, \csc\alpha = \frac{H}{O}$$

To see the representation of a hyperbolic function as the proportion of each side of a right triangle, one might refer to the hyperbolic triangle within a hyperbolic sector (Wares, 2019)

5.3.3. Special Right Triangles

The values of the trigonometric functions can be evaluated exactly for certain angles using right triangles with special angles (Maclear, 1923). Two examples of triangles commonly used in trigonometry are the *30–60–90 triangle* and the *45–45–90 triangle*. The *30–60–90 triangle* allows for the evaluation of trigonometric equations for any number over π/6, while the *45–45–90 triangle* allows for the evaluation of trigonometric equations for any number of π/4 (Yadav, 2017).

- **Kepler Triangle:** A Kepler triangle is a right triangle with edge lengths in geometric progression. The ratio of the edges of a Kepler triangle is linked to the golden ratio.

$$\varphi = \frac{1+\sqrt{5}}{2}$$

and can be written: 1: $\sqrt{\varphi} : \varphi$, or approximately 1: 1.2720196: 1.6180339.

The Kepler Triangle is a geometric configuration named after the German mathematician and astronomer Johannes Kepler. The geometric development observed on the outer edges of this right-angled triangle classifies it as a Kepler triangle (Lenard & Brozinsky, 1994; Szpiro, 2003).

5.3.4. Inequalities

In any right triangle, the diameter of the incircle is less than half the hypotenuse, and it is less than or equal to the hypotenuse times $\left(\sqrt{2}-1\right)$.(2−1).)

In a right triangle, denoted by the sides a, b, and hypotenuse c,

$$c \geq \frac{\sqrt{2}}{2}\left(a+b\right)$$

Equality is only achieved in the case of an isosceles triangle. If the height originating from the hypotenuse is represented as h_c, then:

$$h_c \leq \frac{\sqrt{2}}{4}\left(a+b\right)$$

5.3.5. Thale's Theorem

Thales' theorem states that for every point *A* situated on a circle with diameter BC, except for points *B* and *C*, the triangle ABC formed is a right triangle, with vertex *A* representing the right angle (Patsopoulos & Patronis, 2006) (Figure 5.4).

According to the reverse of a geometric theorem, if a triangle with a right angle is enclosed within a circle, it follows that the length of the hypotenuse will be equivalent to the diameter of the circumference of the circle. A corollary may be drawn from the above information, which states that the total length that defines the hypotenuse is equal to twice the distance between the right angles vertex and the center of the hypotenuse (Ostermann et al., 2012).

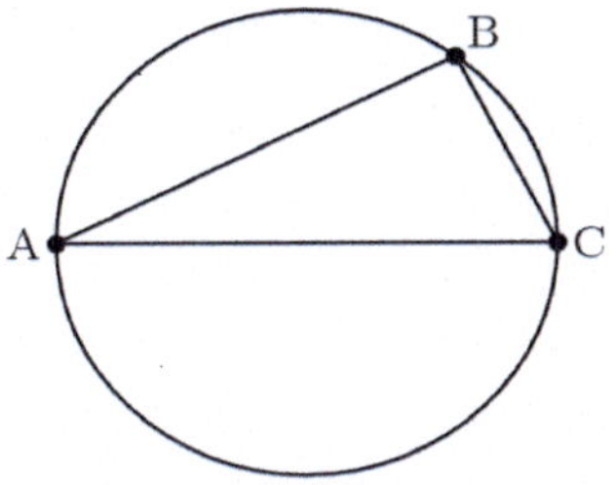

Figure 5.4. A circle intercepting 3 corners of a triangle.

Source: Stack exchange, creative commons.

Moreover, it can be observed that the midpoint of the hypotenuse serves as the exact center of the circle that defines a right triangle. Additionally, the radius of this circle is precisely half of the width of the hypotenuse (Mrabet, 2019).

5.4. MEDIANS

The median of a triangle is a line segment drawn from a vertex to the midpoint of the opposite side of the vertex (Figure 5.5).

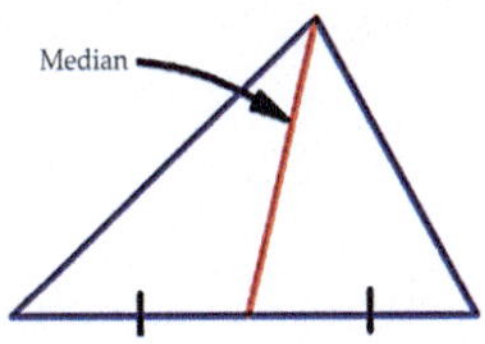

Figure 5.5. Illustration of median and centroid in a triangle.

Source: Varsity tutors, creative commons.

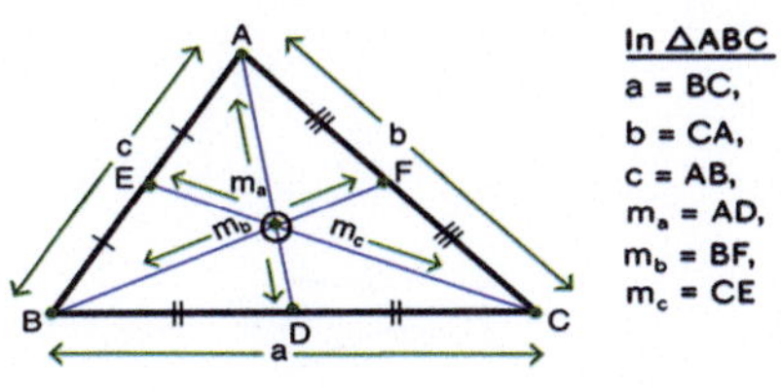

Figure 5.6. Illustration of three medians (m_a, m_b, & m_c) of a triangle along with centroid (O).

Source: Mathmonks, creative commons.

The medians of a triangle are concurrent at a point. The point of concurrency is called the centroid (Figure 5.6).

The formula for the first median of a triangle is as follows:

$$m_a = \frac{\sqrt{(2b^2 + 2c^2 - a^2)}}{2}$$

where the median of the triangle is m_a, the sides of the triangle are a, b, c, and the median is formed on side 'a.'

The formula for the second median of a triangle is as follows:

$$m_b = \frac{\sqrt{(2a^2 + 2c^2 - b^2)}}{2}$$

where the median of the triangle is m_b, the sides of the triangle are a, b, c, and the median is formed on side 'b.'

The formula for the third median of a triangle is as follows:

$$m_c = \frac{\sqrt{(2a^2 + 2b^2 - c^2)}}{2}$$

where the median of the triangle is m_c, the sides of the triangle are a, b, c, and the median is formed on side 'c.'

5.5. EULER LINE

In geometry, the Euler line, named after Leonhard Euler, is a line determined from any triangle that is not equilateral. It is a central line of the triangle, and it passes through several important points determined from the triangle, including the orthocenter, the circumcenter, the centroid, the Exeter point and the center of the nine-point circle of the triangle (Figure 5.7).

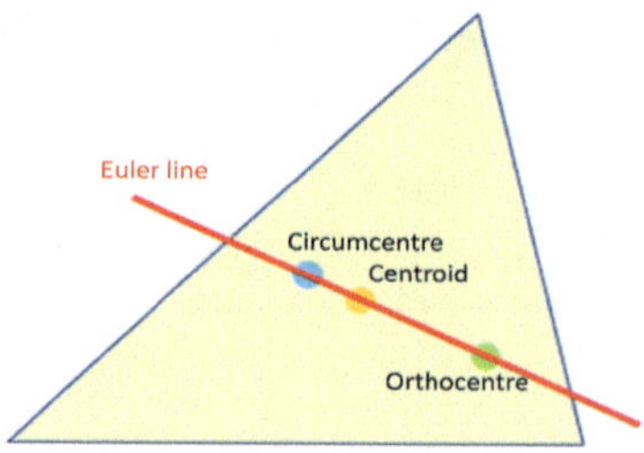

Figure 5.7. Triangle with Euler line.

Source: Open University Mathematics Society, Creative Commons.

5.6. LAW OF SINES

It is important to remember that our triangle definitions of the six trigonometric functions involve ratios of the sides of a right triangle only. These definitions do not apply to oblique triangles. The triangle depicted in Figure 5.8 is classified as oblique; hence, it is not appropriate to assert sin A = a/c. (Whittlesey, 2019; Garg, 2021). Therefore, the Law of Sines and the Law of Cosines are crucial for solving these triangles.

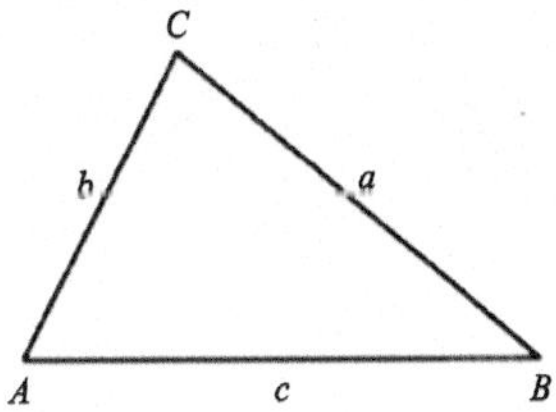

Figure 5.8. Oblique triangle.

Source: Serge, Creative Commons License.

Table 5.2. Solving Oblique Triangles Using Law of Sines

Case	**Method**
AAS *Angle-Angle-Side* A triangle can be determined by specifying both angles and the length of the side opposing one of those angles.	Law of Sines
ASA *Angle-Side-Angle* A triangle can be determined when both angles plus the included side are given.	
SSA *Side-Side-Angle* A triangle can be determined when both sides plus an angle opposing one of each of the sides is given.	

Source: Serge, Creative Commons License.

Each triangle possesses a sum of three sides and three angles. In order to determine the properties of a triangle, it is necessary to possess knowledge of a minimum of three out of the six specified values. Table 5.2 provides a comprehensive summary of the many scenarios that may arise while solving the triangles, along with the corresponding methods employed to tackle each instance (Nguyen & Pham, 2020).

There are many relationships that exist between the sides and angles in a triangle. One such relationship is called the *law of sines*, which states that the ratio of the sine of an angle to the length of the side opposite that angle is constant in any triangle (Balestro et al., 2017).

- **Proof of Law of Sine or Rule of Sine:** The altitude, denoted as h, of the triangle seen in Figure 5.9 can be expressed in relation to either sin A or sin B, according to the specific right triangle under consideration.

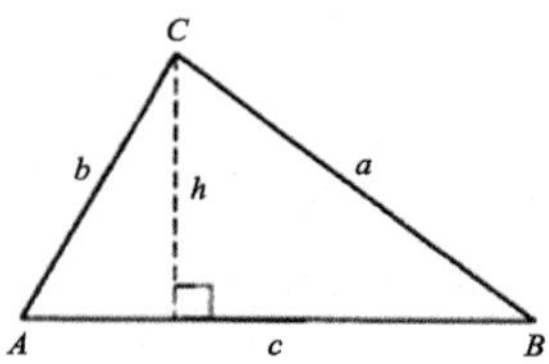

Figure 5.9. ABC triangle.

Source: Serge, Creative Commons License.

$$\sin A = \frac{h}{b} h = b\, sinA$$

$$\sin B = \frac{h}{a} h = a\, sinB$$

Since h is equivalent to itself, we devise;

$$h = h$$

$$b\, sinA = a\, sinB$$

$$b\, sinA \,/\, ab = a\, sinB \,/\, ab\; divide\, both\, sides\, by\, ab$$

$$\frac{sinA}{a} = \frac{sinB}{b} divide\, out\; common\; factors$$

By considering the altitude extending from point A, we may establish a third ratio in the rule of sines, namely sinC, which is equivalent to the two prior ratios.

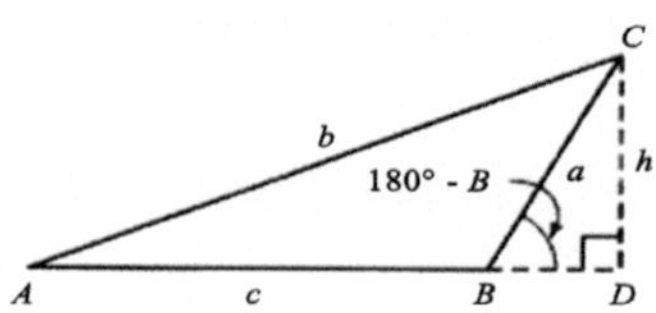

Figure 5.10. Triangle ABCD.

Source: Steven, Creative Commons License.

In triangle BDC, we have;

$$\sin(180° - B) = \frac{h}{a}$$

Then

$$\sin(180° - B) = \sin 180° \ \cos B - \cos 180° \sin B$$

$$= (0)\cos B - (-1)\sin B$$

$$= \sin B$$

Therefore, it can be observed that the sine of angle B may be expressed as the ratio of the length of the side opposite to angle B (Weber et al., 2012) to the length of the hypotenuse (Kimberling, 1994), which aligns with the previously derived outcome. By considering triangle ADC, sin A = h/b. The equations above are identical to the first ones employed during the derivation of the rule of sines governing the acute triangle seen in Figure 5.10. Henceforth, the derivation will align with our prior derivation.

The law of sines can be employed to determine unknown components of triangles in cases when two angles and one side are provided (Wolf & Krotzsch, 1995).

5.6.1. Two Angles and One Side

The previous example presents a scenario where two angles and the side opposing one of the angles are provided. It is worth noting that in the field of geometry, the above-mentioned relationships are required to equate two triangles to establish their congruence by using the AAS Theorem (Heath, 2015).

Example: In triangle ABC, where A=30 °, $B = 70°, and$ a = 8.0cm, find the length of side c.

Solution: We begin by drawing a picture of triangle ABC (it does not have to be accurate) and labeling it with the information we have been given (Figure 5.11):

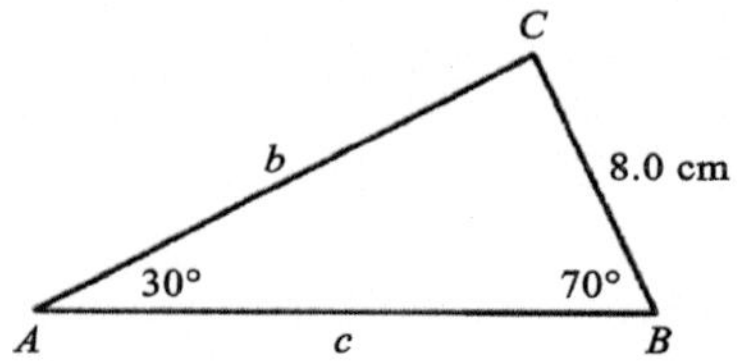

Figure 5.11. Triangle ABC.

Source: Steven, Creative Commons License.

When we use the law of sines, we must have one of the ratios given to us. In this case, since we are given a and A, we have the ratio a/sinA. To solve for c, we need to first find angle C. The sum of the angles in any triangle is 180°, so we have

$$C = 180° - (A + B) = 180° - (30° + 70°) = 80°$$

To find side c, we use the following two ratios given in the law of sines.

$$c / \sin C = a / \sin A$$

To solve for c, we multiply both sides by sin C and substitute.

$$c = a \sin C / \sin A \, \mathit{multiply\, both\, sides\, by} \sin C$$

$$= \frac{8.0 \sin 80°}{\sin 30°} \mathit{substitute\, in\, known\, values}$$

$$= \frac{8.0(0.9848)}{0.5000} \mathit{calculator}$$

$$= 16\mathit{cm\, two\, significant\, digits}$$

5.6.2. The Ambiguous Case

In this section, we will extend the law of sines to solve triangles for which we are given two sides and the angle opposite to one of the given sides (Maybank, 1990).

Example: Find angle B in triangle ABC if a = 2, b 6, and A = 30°.

Solution: By applying the law of sines, we have:

$$\frac{sinB}{b} = \frac{sinA}{a}$$

$\sin B = b \sin A / a$

$= 6 \sin 30° / 2$

$= 1.5$

Because sin B can never be larger than 1, no right-triangle exists for which $a = 2$, $b = 6$, and $A = 30°$. Figure 5.12 illustrates what went wrong here. No matter how we orient side a, we cannot complete a triangle (Figure 5.13).

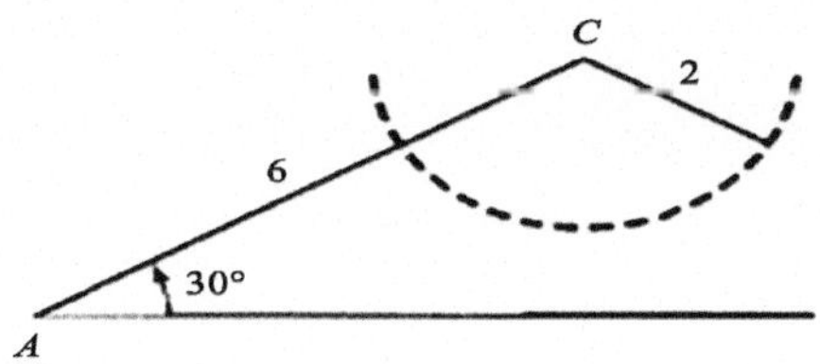

Figure 5.12. Triangle ABC with B angle missing.

Source: William, Creative Commons License.

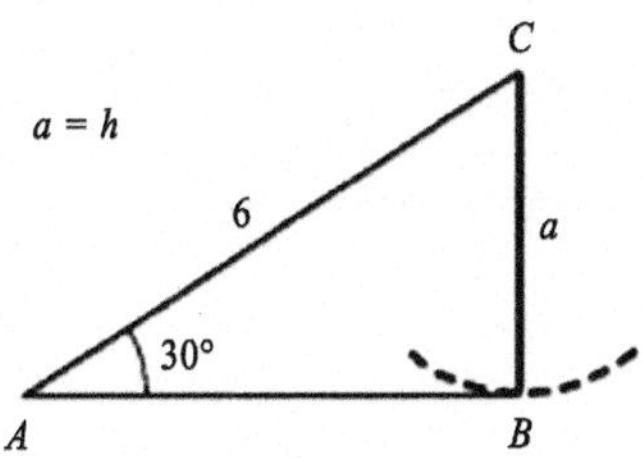

Figure 5.13. Right triangle.

Source: William, Creative Commons License.

5.7. THE LAW OF COSINES

The Law of Cosines, also known as the Cosine Rule, is a fundamental trigonometric formula used in geometry to find the lengths of the sides or the measures of the angles in a non-right (oblique) triangle. It's particularly useful when you have information about the lengths of two sides of a triangle and the included angle (the angle between those two sides), but you don't have information about the right angle.

- The Law of Cosines states:

For a triangle with sides of lengths a, b, and c, and an included angle C (opposite side c), the relationship is as follows:

$$c^2 = a^2 + b^2 - 2ab.\text{Cos}(C)$$

- In this Equation:
  - c is the length of the side opposite angle C.
  - a and b are the lengths of the other two sides.
  - cos(C) is the cosine of angle C.

This formula allows you to find the length of one side (c) when you know the lengths of the other two sides (a) and (b) and the included angle (C). It can also be rearranged to find angles in a triangle when you know the lengths of the sides (Figure 5.14). For example:

$$\cos(C) = \sqrt{(a^2 + b^2 - c^2)} / 2ab$$

Law of Cosines is a powerful tool for solving problems involving triangles of different shapes and sizes.

Here are other variants of the cosine law (Nguyen, 2019) (Table 5.3):

$$a^2 = b^2 + c^2 - 2bc \cos A$$
$$b^2 = c^2 + a^2 - 2ca \cos B$$
$$c^2 = a^2 + b^2 - 2ab \cos C$$

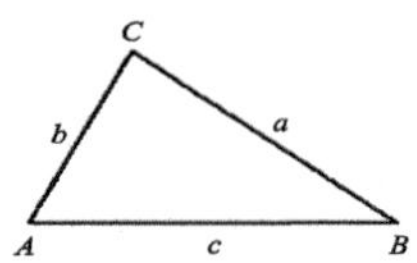

Figure 5.14. Triangle ABC.

Source: Miller, Creative Commons License.

Table 5.3. Solving Oblique Triangles Using Law of Cosine

Case	Method
SAS *Side-Angle-Side* A triangle can be determined when two sides and an angle to them are given.	Law of Cosines
SSS *Side-Side-Side* A triangle may be determined when the lengths of all three sides are known.	

Source: Serge, Creative Commons License.

The law of cosines may be employed to determine the unknown elements of triangles in two scenarios: when we are provided with both sides and an angle formed between them or when all three sides are available.

5.8. TRIGONOMETRIC RATIOS IN TRIANGLES

The fundamental trigonometric ratios play a pivotal role in the deeper understanding of trigonometry, as they facilitate a more profound comprehension of the interconnections between angles and sides inside a triangle (Yigit, 2016).

Trigonometric ratios are defined as precise ratios that establish a relationship between the lengths of the sides of a triangle and the measurements of its angles. The ratios serve as the fundamental components of the functions of trigonometry, including sine (sin), cosine (cos), cosecant (csc), secant (sec), and cotangent (cot). The trigonometric proportions of a specific angle in a right-angled triangle refer to the ratios of the lengths of the triangle's sides in relation to that angle (Sánchez et al., 2023).

5.8.1. Discovering Trigonometric Ratios

Consider a right-angled triangle angled at B (Figure 5.15).

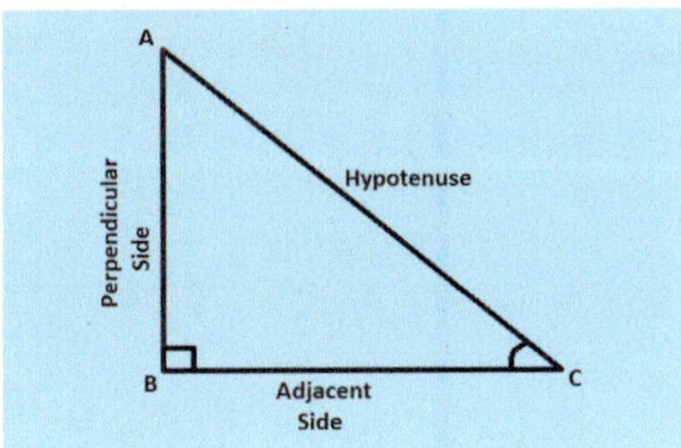

Figure 5.15. Right angled triangle.

Source: John, Creative Commons License.

With respect to ∠C, the trigonometric ratios are defined as follows:

i. The sine of an angle within a triangle is determined by dividing the height of the side opposing the angle by the hypotenuse for a right-angled triangle. The Law of Sines may be utilized to compute the sine ratio in triangles that are not right-angled.

ii. The cosine of an angle in a triangle is defined as the ratio of the length of the adjacent side to the length of the hypotenuse in a right-angled triangle. In non-right-angled triangles, the cosine ratio can be determined using the Law of Cosines.

iii. The tangent of an angle in a triangle is defined as the ratio of the length of the opposite side to the length of the adjacent side in a right-angled triangle. In non-right-angled triangles, the tangent function can be expressed as the ratio of sine and cosine functions.

iv. The cosecant function can be defined as the reciprocal or multiplicative inverse of the sine function.

v. The secant function can be defined as the reciprocal of the cosine function.

vi. The cotangent function can be defined as the reciprocal of the tangential function (Cetin, 2015).

The above ratios are denoted by the abbreviations: sin, cos, tan, cosec, sec, and cot, respectively. In the case of triangle ABC, the proportions are defined as follows:

$$\sin C = (side\, opposite\, to\, \angle C) / (Hypotenuse) = \frac{AB}{AC}$$

$$\cos C = (side\, adjacent\, to\, \angle C) / (Hypotenuse) = \frac{BC}{AC}$$

$$\tan C = \frac{(side\, opposite\, to\, \angle C)}{(side\, adjacent\, to\, \angle C)} = \frac{AB}{BC} = sin\angle C / \cos \angle C$$

$$\text{cosec} C = \frac{1}{sin} \angle C = \frac{(Hypotenuse)}{(side\, opposite\, to\, \angle C)} = \frac{AC}{AB}$$

$$\sec C = \frac{1}{cosC} = \frac{(Hypotenuse)}{(side\, adjacent\, to\, \angle C)} = \frac{AC}{BC}$$

$$\cot C = \frac{1}{tanC} = (side\, adjacent\, to\, \angle C) / (side\, opposite\, to\, \angle C) = \frac{BC}{AB}$$

It can be observed that there are an unlimited number of right triangles satisfying these conditions. However, it can be concluded that the ratios associated with ∠C in all of these triangles will remain constant. This implies that the ratios remain unaffected by variations in the lengths of the triangle's sides.

Trigonometric ratios are of significant importance in right-angled triangles owing to the inherent simplicity of their geometric configuration. Using these ratios enables us to address various difficulties pertaining to right-angled triangles effectively.

5.8.1.1. Applications of Sine, Cosine, and Tangent Ratios

The cosine, tangent, and sine ratios are crucial in the computation of variable angles and side lengths within right-angled triangles. Architects, designers, and physicists employ these ratios to analyze diverse buildings, ascertain distances, and resolve practical difficulties.

5.8.1.2. Special Right-Angled Triangles and Trigonometry

Special right-angled triangles, such as 30–60–90 and 45–45–90 triangles, have specific trigonometric ratios that follow unique patterns. Understanding these patterns facilitates computational processes and improves problem-solving abilities.

5.8.2. Trigonometric Ratios in Non-Right-Angled Triangles

Trigonometric ratios transcend the realm of right-angled triangles and assume a more complicated nature when applied to non-right-angled triangles. The utilization of the aforementioned Law of Sines and the Law of cosines is crucial in the determination of these ratios.

5.8.2.1. The Law of Sines

The Law of Sines establishes a relationship between the sine ratios of the angles and the lengths of the corresponding opposite sides in any triangle. It is a powerful tool for solving non-right-angled triangles when certain information is given.

5.8.2.2. The Law of Cosines

The Law of Cosines enables us to find the length of one side of a triangle when the measures of the other two sides and the included angle are known. It is another potent tool for solving triangles that do not have a right angle.

5.8.3. Applications of Trigonometric Ratios

The practical uses of trigonometric ratios encompass a wide range of fields and disciplines. These ratios, which involve the relationships between the angles and sides of triangles, find utility in several real-world scenarios, for instance:

i. Navigation, surveying, and mapping.

ii. Design of buildings and calculation of forces.

iii. Analysis of vectors, missiles, and circular motion.

iv. Quantification of the spatial separation between celestial entities (Ruthven et al., 2008).

5.9. TRIANGULATION

Triangulation is a basic geometric concept that plays a pivotal role in a wide range of academic fields, including mathematics, physics, navigating, mapping, and computer vision. Triangulation fundamentally operates on the fundamental idea that a triangle may be uniquely determined by any three points that are not aligned in a plane with two dimensions or not coplanar in a three-dimensional space. The notion described serves as the fundamental basis for the technique known as triangulation, which involves utilizing the characteristics of triangles to deduce undisclosed locations, distances, or angles (Luo et al., 2021).

5.9.1. Triangles in Two and Three Dimensions

Triangulation, as applied in the fields of mapping and land surveying, encompasses the process of determining a distance that is unidentified without employing direct measurement methods. This phenomenon is achieved by the utilization of hypothetical triangles. Triangulation proves to be quite advantageous in the context of measuring objects of considerable height. Engaging in the act of ascending a lofty tree with the intention of ascertaining its vertical extent entails a considerable degree of risk. Nevertheless, by distancing oneself from the tree, one might conduct many measurements in order to ascertain its height (Aholt et al., 2012) (Figure 5.16).

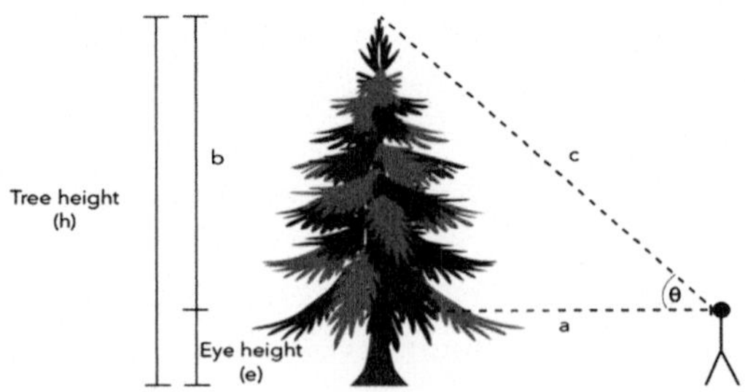

Figure 5.16. Heights and angles involved in tree measurement.

Source: Robert, Creative Commons License.

5.9.1.1. How to Calculate the Height of a Tree?

1. Measure the distance between yourself and the tree. It is denoted as "a," which represents the length of the neighboring edge of the triangle.
2. Find the angular measurement between the height of your eye and the uppermost point of the tree. The symbol shown is denoted as θ. The angle can be determined by utilizing a digital inclinometer or a mobile application designed for measuring inclinations.
3. In order to get the value of b, which represents the opposite side, it is necessary to employ the formula tan 0 = b/a. Subsequently, the cross-multiplication approach may be utilized to solve for b.

$b = tan\theta$

Even if you don't know trigonometry, this task can still be accomplished using a scientific calculator. To get the numerical value of the tangent function, denoted as "tan θ," input the desired angle and afterwards activate the tangent button on your calculator. After the calculation of **b**, it is important to include the vertical distance between the surface of the ground and the observer's eye, denoted as **e**, in order to determine the height of the tree, represented as **h.** Another intriguing discovery pertains to the similarity of two right triangles. In the field of mathematics, two geometric figures are considered comparable if they can be transformed using operations such as reflection, rotation, or dilation to coincide precisely with one another.

Examine the two congruent triangles depicted below. These two entities exhibit similarities, thereby allowing us to assert that:

$\Delta pqr \sim \Delta PQR$

The ~ sign represents similar mathematical things (Figure 5.17):

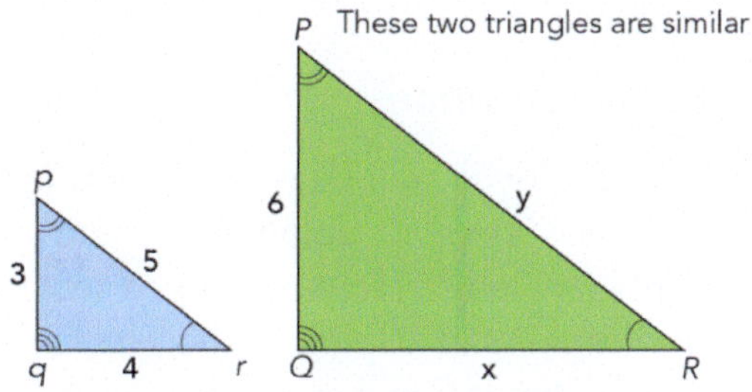

Figure 5.17. Triangle PQR is similar to triangle PQR.

Source: Robert, Creative Commons License.

In order to get the value of the tangent function for a given angle θ, please input the angle and proceed by pressing the letter "t" on your calculator. It is evident that the angles exhibit uniformity, whereas the measured lengths of the sides do not display such consistency. An interesting principle applicable to identical triangles is that their corresponding sides are proportionate. This implies that if we evaluate the sides in the form of fractions, they will exhibit equivalence.

When assigning values to the corresponding sides of each triangle, such as sides pq and PQ, the resulting values are as follows:

$$pq = 3$$

$$PQ = 6$$

The scale factor is the term used to refer to this common ratio. In the present scenario, the factor of scale is determined to be 2.

The concepts of resemblance and scale factor have applicability beyond triangles, encompassing other geometric forms as well.

5.9.2. Control Networks

Triangulation is essential in establishing control networks, which provide a framework for surveying and mapping large regions with high accuracy. These networks consist of precisely surveyed points that serve as reference or control points for subsequent surveys.

Triangulation is also a widely employed technique in the field of particle physics investigations, with the purpose of accurately ascertaining the precise spatial coordinates of particles that are generated within particle colliders. Through the utilization of trajectory measurements, scientists are able to accurately recreate the pathways of particle interactions (Turek & Budzik, 2022).

5.9.2.1. Engineering Applications

Triangulation is widely employed by engineers in a range of applications, such as 3D scanning, measurement, automation, and computer-assisted design (CAD). Triangulation is a widely employed technique in the field of 3D scanning for the purpose of capturing the geometric characteristics of objects existing in three-dimensional space. Similarly, within the field of robotics, triangulation plays a crucial role in ascertaining the precise location and orientation of automated arms or sensors (Voronina et al., 2017).

5.9.2.2. Trigonometric Ratios and Triangles

Trigonometry is a fundamental tool in the field of triangulation, facilitating the determination of angles and lengths in both right-angled and non-right-angled triangles. The tangent, sine, and cosine functions play a crucial role in the computation of angles and distances.

5.9.2.3. Computer Vision and 3D Reconstruction

Triangulation is a widely employed technique in the field of computer vision, wherein the objective is to rebuild three-dimensional (3D) points by leveraging several two-dimensional (2D) pictures captured from distinct views. Through the examination of pixel correspondences between pictures, computer vision algorithms possess the capability to ascertain the three-dimensional location of points inside the observed scene (Fathi et al., 2015).

The process of capturing the geometry of objects in three dimensions with 3D scanners involves the application of triangulation techniques. The scanners utilize a light or laser projection to create a pattern on the surface of the object, which is then analyzed to ascertain 3D coordinates of various locations on the object (Mescheder et al., 2019).

5.9.2.4. Triangulation and Geographic Information Systems (GIS)

GIS systems employ the technique of triangulation to facilitate the gathering, examination, and representation of data across diverse domains such as urban planning, environmental research, and archaeology. Through the integration of spatial information and triangulation-based tests, GIS have the capability to generate intricate maps and conduct comprehensive analysis of geographical patterns (Fabri & Pion, 2009).

5.10. APPLICATIONS OF TRIANGLES IN NAVIGATION

Navigation in triangles refers to using trigonometry and geometry to determine one's position and direction when moving in a two-dimensional or three-dimensional space. This navigation technique is particularly useful when direct measurements are challenging or impossible, such as at sea or in vast and unfamiliar terrains (Mayer et al., 2016). Triangles play a pivotal role in navigation, serving as fundamental tools for determining position, direction, and distance. Navigators, both ancient and modern, rely on triangle-based principles to explore, travel, and reach their destinations efficiently. Here's an overview of how triangles are indispensable in navigation (Sotelo et al., 2020):

5.10.1. Distance Measurement

Triangles provide a simple method to estimate distances between two points when one side of the triangle is known. This principle is the basis for various distance measurement techniques in navigation, such as:

- **Dead Reckoning:** In dead reckoning, mariners calculate their position by estimating how far they've traveled in a straight line from their starting point, using a known speed and time. This is essentially applying the Pythagorean theorem to a triangle formed by time, speed, and distance.
- **Radar Ranging:** Radar systems measure the distance to a target by sending out electromagnetic waves and timing how long it takes for the waves to bounce back. This timing is essentially a distance calculation using the speed of light, forming the basis of radar navigation.

5.10.2. Direction Finding

Triangles are essential for determining direction or bearing. By measuring angles relative to a fixed reference point (usually North), navigators can pinpoint their heading (Figure 5.18):

- **Compass Navigation:** Traditional compasses rely on the magnetic properties of the Earth to indicate the direction of the North Pole. Navigators use compass bearings to set their course.

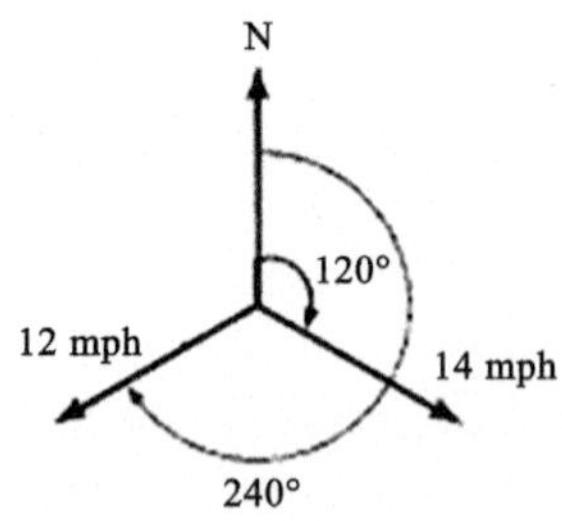

Figure 5.18. Two vectors representing ship velocities.

Source: Stephen, Creative Commons License.

- **Celestial Navigation:** Celestial navigation involves measuring the angles between celestial bodies (like stars, the sun, and the moon) and the horizon. By creating celestial triangles, navigators can calculate their latitude and longitude (Ma et al., 2019).

5.10.3. Position Fixing

Triangulation, a technique extensively used in navigation, involves determining an unknown location by measuring angles to it from known points. It's often employed to fix a ship's position at sea:

GPS relies on a network of satellites orbiting Earth to provide accurate positioning information. Your GPS device calculates your position by triangulating signals from multiple satellites (Sotelo et al., 2015; 2017).

Airplanes and rockets also use trigonometric calculations for navigation. In the context of objects in motion through the atmosphere, the word "airspeed" is employed to denote the velocity of the object with respect to the surrounding air, whereas "ground speed" refers to the velocity of the item in relation to the Earth's surface (Figure 5.19).

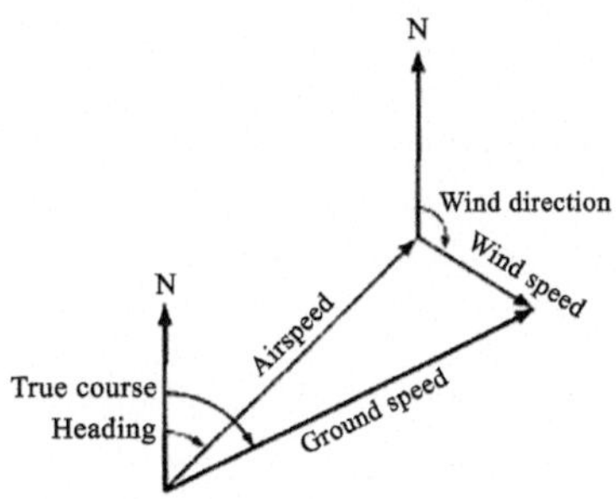

Figure 5.19. Ground speed.

Source: Stephen, Creative Commons License.

5.10.4. Navigational Charts

Nautical charts used for sea and marine navigation are filled with triangles. The triangle symbolizes a magnetic north arrow, indicating the orientation of the chart concerning the magnetic north. This orientation is crucial for aligning the chart with the compass.

5.10.5. Navigational Mathematics

Trigonometry, the branch of mathematics dealing with triangles and their relationships, plays a crucial role in navigation. Concepts like the law of sines and the law of cosines are fundamental for solving various navigation problems, especially in celestial navigation.

5.10.6. Emergency Navigation

In emergency situations where navigational equipment fails, mariners and aviators can use basic triangle principles to estimate their position, plot a course, and navigate to safety.

In essence, triangles are the building blocks of navigation. They enable sailors, pilots, and explorers to calculate distances, determine directions, and fix positions, ensuring safe and accurate navigation across land, sea, and air. Whether it's using basic trigonometric principles or advanced satellite technology, triangles remain at the core of modern navigation systems.

5.11. SUMMARY

This chapter provides an introduction to triangles, where the fundamental nature of three-sided polygons reveals a multitude of mathematical phenomena and practical implications in the physical world. This study endeavors to undertake an expedition of inquiry, revealing the multifaceted nature of triangles and their significant impact on a range of academic fields, including trigonometry, navigation, surveying, and related subjects.

Trigonometry is based on the fundamental concept of right-angled triangles, which are defined by the presence of a 90-degree angle. The Pythagorean Theorem, a perennial mathematical principle, elucidates the correlation among the sides of a triangle with a right angle, stating that the square root of the hypotenuse is equivalent to the combined squares of the remaining two sides. The aforementioned fundamental concept is utilized in several contexts, such as the determination of distances, the computation of unknown sides or angles, and the resolution of practical issues within disciplines such as engineering, architecture, and surveying.

Within the domain of triangles, the Law of Sines and the Law of Cosines emerge as influential instruments for comprehending both right and non-right-angled triangles. The Law of Sines establishes a mathematical link that links the sine of an angle and the dimension of the sides that are opposite to that angle. On the other hand, the Law of Cosines explores the interplay between the lengths of the sides and the measures of the angles in any given triangle. These rules provide scientists, engineers, and navigators with the authority and capability to address intricate spatial issues and acquire significant knowledge about the surrounding environment.

The trigonometric ratios, including sine, cosine, and tangent, play a pivotal role in comprehending the connections between degrees and sides inside triangles. The reciprocals of trigonometric functions, namely cosecant, secant, and cotangent, serve to deepen the comprehension of geometric structures and phenomena. Trigonometric principles are utilized by scientists and engineers to get valuable insights into a wide range of physical processes, rendering them essential instruments in different applications.

Triangulation, an exceptional geometric methodology, arises as a potent instrument in the fields of surveying, navigation, and position systems. The use of triangulation by land surveyors is a proficient technique for the measurement of distances and angles, facilitating the accurate cartography of terrains and the delineation of land demarcations. The practice of celestial navigation is predicated upon the utilization of triangulation as a means to ascertain the precise locations of seafarers while traversing the vast expanses of the ocean, therefore unraveling the enigmatic nature of both latitude and longitude. Furthermore, the Global locating System (GPS) employs a method of satellite triangulation, therefore transforming worldwide navigation and enhancing the precision of locating.

Upon the culmination of the chapter titled "Triangles," one is left in a state of profound admiration for the significant influence that triangles and trigonometry exert on our global landscape. Throughout the course of human history, triangles have played a pivotal role in various aspects of our civilization. In ancient times, they served as the foundation for land measuring and navigation techniques, enabling early civilizations to navigate and explore their surroundings. In contemporary times, triangles continue to have significance as they facilitate the functioning of cutting-edge technologies like GPS, robotics, and computer graphics. These innovations have greatly contributed to our understanding and investigation of the physical universe. The convergence of mathematics and reality is observed, wherein the fundamental nature of triangles is unveiled as both refined conceptual constructs and utilitarian instruments. The enduring impact of triangles and trigonometry serves as a source of inspiration for successive cohorts, driving us into novel realms of inquiry and advancement. The ongoing role of geometry in molding our knowledge of the universe and driving development is exemplified by the ageless grace of triangles.

REFERENCES

1. Abdukadirovich, S. U., & Abduganievich, D. U., (2020). About the issues of geometrical inequalities and the methods of their solution. *European Science, 7*(56), 5–10.
2. Aholt, C., Agarwal, S., & Thomas, R., (2012). *A QCQP Approach to Triangulation.* Paper Presented at the Computer Vision–ECCV 2012: 12th European conference on computer VISION, Florence, Italy, October 7-13, 2012, Proceedings, Part I 12.
3. Ann, P. T., (1988). *The Right-Angled Triangle* (Vol. 50, pp. 200–300). Presidential Address Delivered to the Singapore Mathematical Society; Department of Mathematics, National University of Singapore: Singapore.
4. Balestro, V., Martini, H., & Teixeira, R., (2017). Geometric properties of a sine function extendable to arbitrary normed planes. *Monatshefte für Mathematik, 182*, 781–800.
5. Bayu, C. P., Yohanie, D. D., & Handayani, A. D., (2021). Analysis of student errors in solving comparative trigonometry problems of right-angled triangles based on watson's criteria. *Jurnal Math Educator Nusantara: Wahana Publikasi Karya Tulis Ilmiah di Bidang Pendidikan Matematika, 7*(2), 161–171.
6. Bóna, M., & Tóth, G., (1996). A Ramsey-type problem on right-angled triangles in space. *Discrete Mathematics, 150*(1–3), 61–67.
7. Bulut, A., (2012). *Investigating Perceptions of Preservice Mathematics Teachers on their Technological Pedagogical Content Knowledge (TPACK) Regarding Geometry* (No. 10, pp. 20–30). Middle East Technical University.
8. Cetin, O. F., (2015). Students perceptions and development of conceptual understanding regarding trigonometry and trigonometric function. *Educational Research and Reviews, 10*(3), 338–350.
9. Chappell, J. M., Drake, S. P., Seidel, C. L., Gunn, L. J., Iqbal, A., Allison, A., & Abbott, D., (2014). Geometric algebra for electrical and electronic engineers. *Proceedings of the IEEE, 102*(9), 1340–1363.
10. Clarke, R. J., (2005). 89.21 Incircles of right-angled triangles. *The Mathematical Gazette, 89*(514), 82–86.
11. Culver, J., Choe, R., Holboke, M., Zubkov, L., Durduran, T., Slemp, A., . Yodh, A., (2003). Three-dimensional diffuse optical tomography in the parallel plane transmission geometry: Evaluation of a hybrid

frequency domain/continuous wave clinical system for breast imaging. *Medical Physics, 30*(2), 235–247.

12. Curtis, R. A., Evans, D. L., Rathmell, J. G., & Rose, R. J. (1997, October). Use of global positioning systems (GPS) in data acquisition for equine exercise physiology. In *Proceedings of the 19th Annual International Conference of the IEEE Engineering in Medicine and Biology Society.'Magnificent Milestones and Emerging Opportunities in Medical Engineering'(Cat. No. 97CH36136)* (Vol. 4, pp. 1821-1823). IEEE.
13. Darvasi, G., (2003). A unique property of right-angled triangles. *The Mathematical Gazette, 87*(508), 51–59.
14. Epstein, L., & Jacobi, T., (2008). Super medians. *Stan. L. Rev., 61*, 37.
15. Fabri, A., & Pion, S., (2009). *CGAL: The Computational Geometry Algorithms Library* (Vol. 14, pp. 67–100). Paper presented at the Proceedings of the 17th ACM SIGSPATIAL international conference on advances in geographic information systems.
16. Fathi, H., Dai, F., & Lourakis, M., (2015). Automated as-built 3D reconstruction of civil infrastructure using computer vision: Achievements, opportunities, and challenges. *Advanced Engineering Informatics, 29*(2), 149–161.
17. Gao, H., Sun, L., & Wang, J. X., (2021). PhyGeoNet: Physics-informed geometry-adaptive convolutional neural networks for solving parameterized steady-state PDEs on irregular domain. *Journal of Computational Physics, 428*, 110079.
18. Garg, H., (2021). Sine trigonometric operational laws and its based pythagorean fuzzy aggregation operators for group decision-making process. *Artificial Intelligence Review, 54*(6), 4421–4447.
19. Gel□fand, I. M., Saul, M., Gelfand, I., & Saul, M., (2001). *Trigonometry* (Vol. 9, pp. 30–60). Springer.
20. Goodchild, S., & Sriraman, B., (2012). Revisiting the didactic triangle: From the particular to the general. *Zdm, 44*, 581–585.
21. Heath, T., (2015). *Mathematics in Aristotle* (Vol. 8, pp. 200–450). Routledge.
22. Herbst, P., & Chazan, D., (2012). On the instructional triangle and sources of justification for actions in mathematics teaching. *ZDM, 44*, 601–612.

23. Indelman, V., Gurfil, P., Rivlin, E., & Rotstein, H., (2012). Real-time vision-aided localization and navigation based on three-view geometry. *IEEE Transactions on Aerospace and Electronic Systems, 48*(3), 2239–2259.
24. Jumaniyozova, N. K., (2020). Triangle and its striking points. *Theoretical & Applied Science*, 1*(8)*, 50–53.
25. Khan, S. A., (2020). Trigonometric ratios using geometric methods. *Advances in Mathematics: Scientific Journal, 9*(10), 8685–8702.
26. Kimberling, C., (1994). Central points and central lines in the plane of a triangle. *Mathematics Magazine, 67*(3), 163–187.
27. Koniaris, C., Chatterjee, S., & Kleijn, W. B., (2010). *Selecting Static and Dynamic Features Using an Advanced Auditory Model for Speech Recognition* (Vol. 70, No. 3, pp. 70–110). Paper presented at the 2010 IEEE International Conference on Acoustics, Speech and Signal Processing.
28. Lenard, A., & Brozinsky, M. K., (1994). Kepler orbits more geometrico. *The College Mathematics Journal, 25*(2), 90–98.
29. Li, B. L., (2000). Fractal geometry applications in description and analysis of patch patterns and patch dynamics. *Ecological Modeling, 132*(1, 2), 33–50.
30. LINE, S., (2002). *Higher Mathematics, 12,* 45–90.
31. Luo, Y., Mi, Z., & Tao, W., (2021). *Deepdt: Learning Geometry from Delaunay Triangulation for Surface Reconstruction* (Vol. 15, pp. 90–180). Paper presented at the Proceedings of the AAAI Conference on Artificial Intelligence.
32. Ma, X., Ning, X., Chen, X., & Liu, J., (2019). Geometric coplanar constraints-aided autonomous celestial navigation for spacecraft in deep space exploration. *IEEE Access, 7*, 112424–112434.
33. Maclear, M., (1923). Mathematics in current literature. *The Pedagogical Seminary, 30*(1), 48–50.
34. Maybank, S., (1990). The projective geometry of ambiguous surfaces. *Philosophical Transactions: Physical Sciences and Engineering, 17*, 1–47.
35. Mayer, U., Pecchia, T., Bingman, V. P., Flore, M., & Vallortigara, G., (2016). Hippocampus and medial striatum dissociation during

goal navigation by geometry or features in the domestic chick: An immediate early gene study. *Hippocampus, 26*(1), 27–40.

36. Mescheder, L., Oechsle, M., Niemeyer, M., Nowozin, S., & Geiger, A., (2019). *Occupancy Networks: Learning 3d Reconstruction in Function Space* (Vol. 6, pp. 30–90). Paper presented at the Proceedings of the IEEE/CVF conference on computer vision and pattern recognition.
37. Moldavanov, A., (2017). *Classical Right-Angled Triangles and the Golden Ratio* (Vol. 9, pp. 225–660). Paper presented at the Forum Geometricorum.
38. Morales-López, Y., (2019). *Knowledge Evidenced by Prospective Mathematics Teachers When Performing a Task Involving Geometry, Teaching and the Use of Technology, 7,* 67–90.
39. Mosia, M., (2016). *Improving Teachers' Technological Pedagogical Content Knowledge for Teaching Euclidean Geometry Using Integrated Information Communication Technologies Software* (Vol. 3, pp. 65–115). University of the Free State.
40. Moutsios-Rentzos, A., Spyrou, P., & Peteinara, A., (2014). The objectification of the right-angled triangle in the teaching of the Pythagorean theorem: An empirical investigation. *Educational Studies in Mathematics, 85*, 29–51.
41. Mrabet, S., (2019). *The Development of Thales Theorem Throughout History* (Vol. 98, pp. 100–233). Paper presented at the Eleventh Congress of the European Society for Research in Mathematics Education.
42. Nader, H., (2021). Use of the theory of Fischbein and the theory of Shulman for the study of teachers' algorithmic knowledge concerning the concept of the altitude of a triangle. *Journal for the Mathematics Education and Teaching Practices, 2*(2), 71–80.
43. Nguyen, N. G., & Pham, H. T., (2020). Constructivist learning and the law of sines in advanced 10th grade geometry textbooks in Vietnam. *International Journal of Learning, Teaching and Educational Research, 19*(1), 38–59.
44. Nguyen, N. G., (2019). Teaching the law of cosines in advanced 10th grade geometry textbook by way of problem-based learning in Vietnam. *International Journal of Learning, Teaching and Educational Research, 18*(11), 397–413.
45. Nipper, K., & Sztajn, P., (2008). Expanding the instructional triangle:

Conceptualizing mathematics teacher development. *Journal of Mathematics Teacher Education, 11*, 333–341.

46. Nomura, T., Katoh, M., Yamamoto, Y., Miyachi, Y., & Kabashima, K., (2016). Eosinophilic pustular folliculitis: A proposal of diagnostic and therapeutic algorithms. *The Journal of Dermatology, 43*(11), 1301–1306.
47. Okumura, H., (1990). 74.31 incircles and excircles of right-angled triangles. *The Mathematical Gazette, 74*(469), 278–279.
48. Ostermann, A., Wanner, G., Ostermann, A., & Wanner, G., (2012). Thales and pythagoras. *Geometry by Its History*, 3–26.
49. Pan, Z., Kotani, K., & Ohmi, T., (2004). An improved full-search-equivalent vector quantization method using the law of cosines. *IEEE Signal Processing Letters, 11*(2), 247–250.
50. Panigrahi, N., (2014). *Computing in Geographic Information Systems* (Vol. 10, pp. 347–600). CRC Press.
51. Patsopoulos, D., & Patronis, T., (2006). The theorem of thales: A study of the naming of theorems in school. *International Journal for the History of Mathematics Education, 1*(1), 1–20.
52. Rababa, H. A. A., (2022). New rule for finding the area of a right-angled triangle. *International Journal of Advance Scientific Research, 2*(3), 51–58.
53. Ramachandran, A., (2016). Triangle centers and homogeneous coordinates, Part I-trilinear coordinates. *At Right Angles, 5*(1), 40–43.
54. Ruthven, K., Hennessy, S., & Deaney, R., (2008). Constructions of dynamic geometry: A study of the interpretative flexibility of educational software in classroom practice. *Computers & Education, 51*(1), 297–317.
55. Sakarovitch, J., (1995). The teaching of stereotomy in engineering schools in France in the XVIIIth and XIXth centuries: An application of geometry, an "applied geometry," or a construction technique? In: *Entre Mécanique et Architecture/Between Mechanics and Architecture*, (Vol. 5, pp. 205–220). Springer.
56. Sánchez, M. A., Gómez, C. P., Del Rincón, T. O., & Musonda, T. M., (2023). Designing a theoretical proposal using problem–based learning to improve learning of trigonometric ratios among grade eleven students. *American Journal of Educational Research, 11*(2), 53–78.

57. Sandipan, D. S., (2016) *Area Relation of Two Right Angled Triangle in Trigonometric Form, 3,* 67–78.

58. Selvi, H. Z., Oztug, B. I., & Yerci, M., (2010). Triangulation method for area-line geometry-type changes in map generalisation. *The Cartographic Journal, 47*(2), 157–163.

59. ŞENGÜN, K. Ç., & YILMAZ, S., (2021). Examining the efforts of middle school 7th grade students to draw altitude in parallelogram and triangle. *Osmangazi Journal of Educational Research, 8*(1), 220–238.

60. Shakouri, A., Azad, M. H., & Assadian, N., (2018). Orbit estimation using a horizon detector in the presence of uncertain celestial body rotation and geometry. *Acta Astronautica, 148*, 82–88.

61. Song, M., & Yuan, Y., (2008). *Autonomous Navigation for Deep Spacecraft Based on Celestial Objects* (Vol. 8, pp. 35–90). Paper presented at the 2008 2nd International Symposium on Systems and Control in Aerospace and Astronautics.

62. Sotelo, M. I., Alcalá, J. A., Bingman, V. P., & Muzio, R. N., (2020). On the transfer of spatial learning between geometrically different shaped environments in the terrestrial toad, *Rhinella arenarum. Animal Cognition, 23*(1), 55–70.

63. Sotelo, M. I., Bingman, V. P., & Muzio, R. N., (2015). Goal orientation by geometric and feature cues: Spatial learning in the terrestrial toad *Rhinella arenarum. Animal Cognition, 18*, 315–323.

64. Sotelo, M. I., Daneri, M. F., Bingman, V. P., & Muzio, R. N., (2017). Telencephalic neuronal activation associated with spatial memory in the terrestrial toad *Rhinella arenarum*: Participation of the medial pallium during navigation by geometry. *Brain Behavior and Evolution, 88*(3, 4), 149–160.

65. Stillwell, J., & Stillwell, J., (1989). *Mathematics and its History* (Vol. 3, pp. 4–90). Springer.

66. Szpiro, G., (2003). *Kepler's Conjecture: How Some of the Greatest Minds in History Helped Solve one of the Oldest Math Problems in the World* (Vol. 13, pp. 5–19).

67. Tsamir, P., Tirosh, D., & Levenson, E., (2008). Intuitive nonexamples: The case of triangles. *Educational Studies in Mathematics, 69*, 81–95.

68. Turek, P., & Budzik, G., (2022). Development of a procedure for increasing the accuracy of the reconstruction and triangulation process of the cranial vault geometry for additive manufacturing. *Facta*

Universitatis, Series: Mechanical Engineering, 50, 110–229.

69. Ulusoy, F., (2021). Prospective early childhood and elementary school mathematics teachers' concept images and concept definitions of triangles. *International Journal of Science and Mathematics Education, 19*(5), 1057–1078.
70. Voronina, M. V., Moroz, O. N., Sudarikov, A. E., Rakhimzhanova, M. B., & Muratbakeev, E. K., (2017). Systematic review and results of the experiment of a flipped learning model for the courses of descriptive geometry, engineering and computer graphics, computer geometry. *Eurasia Journal of Mathematics, Science and Technology Education, 13*(8), 4831–4845.
71. Wares, A., (2019). Paper folding and trigonometric ratios. *International Journal of Mathematical Education in Science and Technology, 50*(4), 636–641.
72. Weber, E., Tallman, M., Byerley, C., & Thompson, P., (2012). Introducing derivative via the calculus triangle. *Mathematics Teacher, 104*(4), 274–278.
73. Weibel, R., (1993). *On the Integration of Digital Terrain and Surface Modeling into Geographic Information Systems* (Vol. 8, pp. 776–999). Paper presented at the AUTOCARTO-Conference.
74. Whittlesey, M., (2019). *Spherical Geometry and its Applications* (Vol. 113, pp. 78–89). Chapman and Hall/CRC.
75. Wise, D. T., (2012). *From Riches to Raags: 3-Manifolds, Right-Angled Artin Groups, and Cubical Geometry: 3-manifolds, Right-angled Artin Groups, and Cubical Geometry* (Vol. 117, pp. 115–555)
76. Wolf, K. B., & Krotzsch, G., (1995). Geometry and dynamics in refracting systems. *European Journal of Physics, 16*(1), 14.
77. Yadav, D. K., (2017). Exact definition of mathematics. *International Research Journal of Mathematics, Engineering and IT, 4*(1), 34–42.
78. Yi, J. A., Yoo, J. G., & Lee, K. H., (2013). Toward students' full understanding of trigonometric ratios. *Research in Mathematical Education, 17*(1), 63–78.
79. Yigit, K. M., (2016). Mathematics education graduate students' understanding of trigonometric ratios. *International Journal of Mathematical Education in Science and Technology, 47*(7), 1028–1047.

80. Yim, Y., & Cayford, R., (2001). *Investigation of Vehicles as Probes Using Global Positioning System and Cellular Phone Tracking: Field Operational Test*, *5*, 65–78.
81. Yiu, P., (2001). *Introduction to the Geometry of the Triangle* (Vol. 5, pp. 11–66). In: Florida Atlantic University Lecture Notes.
82. Zhang, L., Zhou, Y., & Wijaya, T., (2020). *Hawgent Dynamic Mathematics Software to Improve Problem-Solving Ability in Teaching Triangles* (Vol. 17, pp. 113–233). Paper presented at the Journal of Physics: Conference Series.

CHAPTER 6

3D Trigonometry and Angles

Contents

6.1. INTRODUCTION

This chapter delves into the foundations that extend beyond traditional Euclidean boundaries. The transition from two dimensions to three introduces a shift that challenges our intuitions. Three-dimensional angles encompass spheres, arcs, and solid angles, each layer contributing to a deeper understanding of spatial relationships. A unit sphere is the central point of this chapter. It is a fundamental building block that allows us to measure and comprehend angles of any magnitude within a three-dimensional context.

However, the impact of three-dimensional trigonometry is not limited to theoretical understanding alone. Its applications reverberate across various fields. In astronomy, the precise calculation of angles enables astronomers to locate celestial objects with accuracy. Navigation systems, including GPS, rely on these principles to guide us through complex terrains. Architects and engineers employ trigonometric principles to calculate dimensions and forces, ensuring structures withstand the challenges posed by nature.

6.2. SPHERICAL COORDINATES

In the field of mathematics and geometry, coordinates are essential to find locations in space (Richter, 2007). While Cartesian coordinates (x, y, z) are the go-to system for describing points in three-dimensional space, there are a few instances where a different approach offers more insight. This is where spherical coordinates come into play. Imagine standing at a point and looking out into space – spherical coordinates provide a means to describe where you're looking and how far away the objects are (Suárez et al., 2022).

Spherical coordinates consist of three components that collectively identify a point in three-dimensional space (Schermerhorn & Thompson, 2019).

These components are:

- ρ (rho): The symbol ρ, known as rho, denotes the radial distance or length from the origin to the specific point under consideration.
- θ (theta): The symbol θ denotes the angle formed between the positive x-axis and the line connecting the origin and the point projected onto the xy-plane.
- φ (phi): Phi (φ) is a symbol that denotes the angle formed between the positive z-axis and the line connecting the origin and a given point.

By combining these three values: ρ, θ, and φ, we can specify any point in a three-dimensional space uniquely (Buckmire, 2003).

6.2.1. Visualizing Spherical Coordinates

Imagine standing at the center of a sphere. The radius of the sphere (ρ) defines your distance from the center of the sphere to the point you're interested in. θ gives you the azimuthal angle, which is the angle you need to rotate from the positive x-axis to reach your line of sight. φ represents the polar angle, indicating how high or low you need to tilt your gaze from the positive z-axis.

6.2.2. Converting Spherical Coordinates

Converting spherical coordinates (ρ, θ, φ) to Cartesian coordinates (x, y, z) and vice versa involves trigonometric calculations. The relationships are as follows:

$$x = \rho \times \sin(\varphi) \times \cos(\theta)$$

$$y = \rho \times \sin(\varphi) \times \sin(\theta)$$

$$z = \rho \times \cos(\varphi)$$

These conversions enable us to bridge the gap between the familiar Cartesian system and the spherical realm (Apriandi & Krisdiana, 2016; Schermerhorn & Thompson, 2019).

6.2.3. Steps for Using Trigonometry in 3D

Trigonometry is a mathematical discipline concerned with the interconnections between sides and angles of triangles (Allen & Kupfer, 2000). In three-dimensional space, trigonometry becomes even more significant, allowing us to calculate distances, angles, and various geometric properties. Understanding how to use trigonometry in 3D is essential in fields such as physics, engineering, computer graphics, and navigation (Mahmud et al., 2023).

To use trigonometry in three dimensions, follow these steps:

- **Define the Problem:** Clearly state the problem you need to solve in 3D space. Understand what information is given and what you need to find. Identify the objects, points, and angles involved.

- **Coordinate System:** Establish a 3D Cartesian coordinate system with x, y, and z axes. This will help you represent the positions of points and vectors in 3D space.
- **Identify the Triangles:** In 3D problems, you often work with triangles formed by vectors or lines. Identify the triangles in your problem. For example, you may have a triangle formed by two vectors and a line or a triangle formed by three points in space.
- **Vector Calculations:** If vectors are involved, perform vector operations as needed. This can include vector addition, subtraction, dot products, and cross products. These operations can help you find angles between vectors and calculate distances.
- **Angle Calculations:** Use trigonometric functions (sine, cosine, tangent) to calculate angles between lines, vectors, or planes. For example, if you have two vectors **A** and **B**, you can find the angle θ between them using the dot product:

$$\cos(\theta) = \left(\vec{A}.\vec{B}\left|\vec{A}\right|\left|\vec{B}\right|\right)$$

You can then use the inverse cosine function (arccos) to find θ.

- **Distance Calculations:** To find distances between points or along lines in 3D space, use the 3D distance formula, which is essentially an extension of the Pythagorean theorem:

$$d = \sqrt{(x_2 - x_1)^2 + (y_2 - y_1)^2 + (z_2 - z_1)^2}$$

  Where (x_1, y_1, z_1) and (x_2, y_2, z_2) are the coordinates of the two points.

- **Solving for Unknowns:** Apply the trigonometric and geometric relationships you've identified to solve for the unknown quantities in your problem. Be sure to use the correct units and coordinate system.
- **Check for Ambiguities:** In 3D trigonometry, there can be multiple solutions or ambiguities, especially when dealing with angles and orientations. Ensure that your solution aligns with the context of the problem.
- **Verify Results:** Double-check your calculations and results to ensure they make sense in the context of the problem. Use visualization tools or software to visualize the 3D space and

confirm your findings.

- **Communicate the Solution:** Present your solution clearly, including the values of angles, distances, or any other relevant quantities. Use proper notation and units.

These steps should help you apply trigonometry effectively in three-dimensional space to solve various problems related to geometry, physics, engineering, and more.

6.3. PYRAMID PROBLEMS

The pyramid diagram works for all kinds of problems involving pyramids with square bases.

Example: The pyramid's square base is denoted as ABCD, with each side measuring 5 cm. The pyramid vertex (V) is located 6 cm above the center (O) of the square base. Calculate the angle AVC (Figure 6.1).

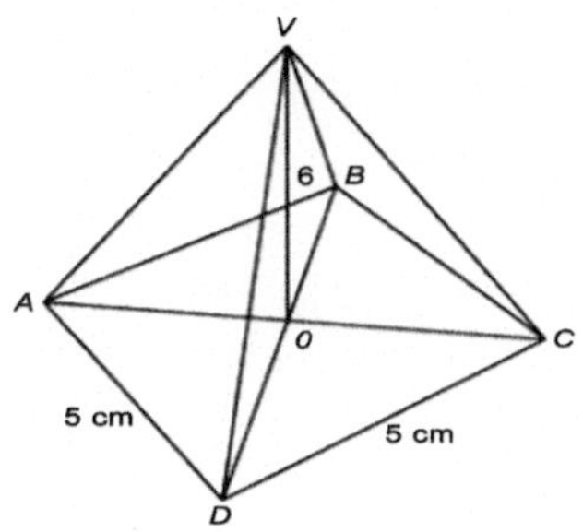

Figure 6.1. ABCD pyramid.

Source: James, Creative Commons License.

There are a number of features you should notice about this diagram:

- The size of the diagram is sufficient to prevent overlapping of points.
- The vertex (V) should be positioned above the center of the base. The recommended approach is initially drawing the base, then drawing the diagonals that connect at point O. Subsequently, put point V vertically above point O, ensuring that the edge VC of the pyramid is not visually aligned with point B.
- It should be noted that the dimensions have not been included in the measurements in cases where their inclusion would result in clutter.

- To solve the problem, you must develop a strategy that involves creating or recognizing right-angled triangles.
- The angle AVC is the vertical angle of the isosceles triangle AVC that is bisected by VO. If you can find the length AO, you can calculate the angle AVC. You can find AO by using the fact that ABCD is a square (Figure 6.2).

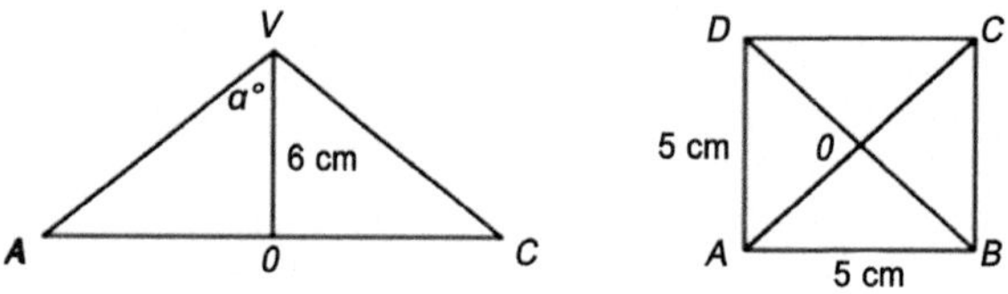

Figure 6.2. ABCD square.

Source: James, Creative Commons License.

Find AC by using Pythagoras's theorem.

$$AC^2 = 5^2 + 5^2 = 50$$

So

$$AC = \sqrt{50cm}\ and\ AO = \frac{1}{2}\sqrt{50}cm$$

Let angle AVO=a °

Then

$$\tan a° = \frac{\frac{1}{2\sqrt{50}}}{6}$$

And

$a° = 30.5089$

Therefore

$angle\ AVC = 2a° = 61.02°$

If you find that you can solve the problem without drawing the subsidiary diagrams in Figure 6.2, then do so. But many people find that it helps to see what is happening. All square-based pyramid problems can be solved using this diagram.

- **Example:** Consider the pyramid PQRSV, whose vertex (V) is positioned symmetrically above the rectangular base PQRS with mid-point O. The measurements of line segments PS and RS are 6 cm and 5 cm, respectively. In comparison, the slant height VP is 8 cm. Determine the angle formed by the edge VP concerning the ground.

The condition is depicted in Figure 6.3.

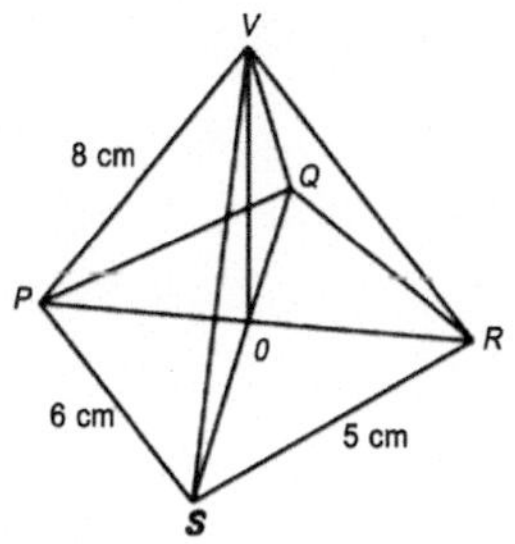

Figure 6.3. PQRSV pyramid.

Source: Andries, Creative Commons License.

Let the required angle VPO be α°. The value of α° in triangle VPO may be determined if the PO or VO lengths are known. The point of origin may be determined from the rectangle PQRS (Rogers & Pope, 2016) (Figure 6.4).

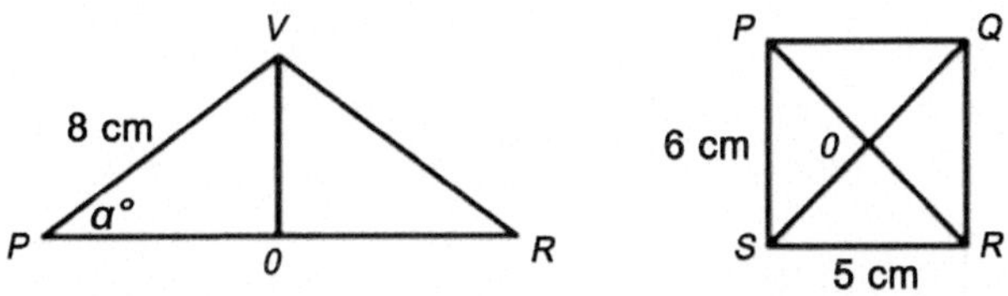

Figure 6.4. Rectangle PQRS.

Source: Andries, Creative Commons License.

Using Pythagoras's theorem gives:

$$PR = \sqrt{PS^2 + RS^2} = \sqrt{6^2 + 5^2} = \sqrt{61}$$

thus,

$$PO = \frac{1}{2}\sqrt{61cm}$$

Then

$$\cos\alpha° = \frac{PO}{VP} = \frac{\frac{1}{2}\sqrt{61}}{8}$$

$$a° = 60.781$$

Hence, the vertex VP forms an angle of roughly 60.78 degrees relative to the horizontal axis.

6.4. BOX PROBLEMS

Now, let's discuss box problems involving 3D angles (Figure 6.5).

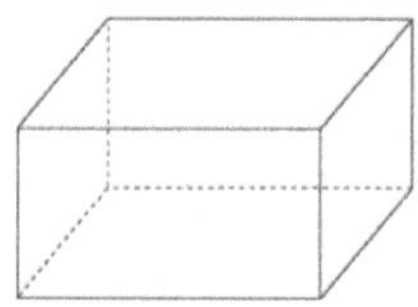

Figure 6.5. A 3D box.

Source: Hugh, Creative Commons License.

The conventional method for drawing a box involves the creation of two similar rectangles, with one positioned in the background relative to the other. Subsequently, the corners are connected in an acceptable manner for the desired outcome. Employing a dotted line technique might prove advantageous to enhance visual clarity and disambiguate the positioning of different faces.

- **Example:** A room has a length of 4 m, width 3 m and height 2 m. Find the angle that a diagonal of the room makes with the floor (Figure 6.6).

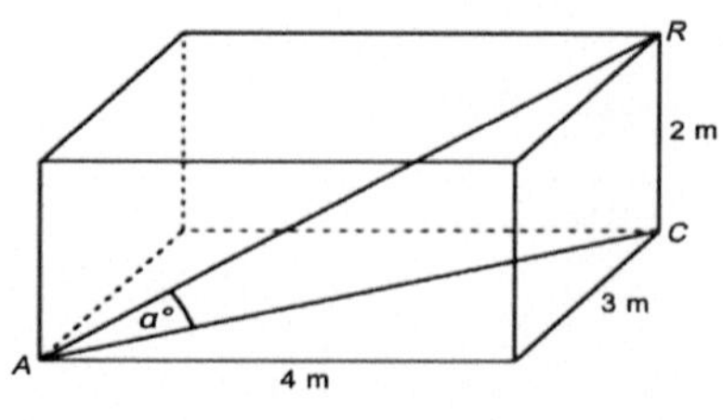

Figure 6.6. Box showing room length, width, and height.

Source: Hugh, Creative Commons License.

Let the diagonal of the room be AR, and let AC be the diagonal of the floor. Suppose the required angle is α°. You can use Pythagoras's theorem to find the diagonal AC of the floor and subsequently find the angle α°.

$$AC = \sqrt{3^2 + 4^2} = \sqrt{25} = 5$$

So

$$tana° = \frac{2}{5}$$

And

$$a = 21.8014$$

Hence, the diagonal forms an angle of roughly 21.80° relative to the floor.

- **Example:** The top, P, of a pylon situated on a flat surface form an angular measurement of 10° at a specific location S, 50 meters directly south, and 5° at a location W, situated west of the pylon. Determine the precise distance in a southwesterly direction, rounded to the nearest meter (Figure 6.7).

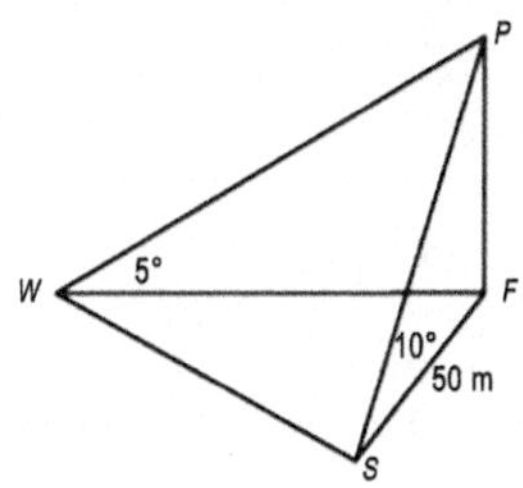

Figure 6.7. Pylon standing on ground level.

Source: James, Creative Commons License.

This illustration is a part of the box diagram seen in Figure 6.5, with certain lines omitted. The foot of the pylon is denoted by the symbol F. To determine the value of SW, it is necessary first to ascertain the vertical dimension of the pylon. Subsequently, the length of FW should be determined, followed by using Pythagoras theorem to compute the value of SW.

In triangle PFS,

$$tan10° = \frac{PF}{50}, so\ PF = 50tan10°$$

In Triangle PFW,

$$tan5° = \frac{PF}{FW} = \frac{50tan10°}{FW}, FW = 50tan10° / tan5°$$

Finally,

$$SW^2 = FW^2 + SF^2 = \left(\frac{50tan10°}{\tan 5°}\right)^2 + 50^2$$

The value of SW is 112 units, rounded to the nearest meter.

6.5. WEDGE PROBLEMS

This section discusses the problems involving a wedge. The wedge is only part of a box, so you could think of a wedge problem as a special case of a box problem, but it is easier to think of it as a separate category (Qi et al., 2012) (Figure 6.8).

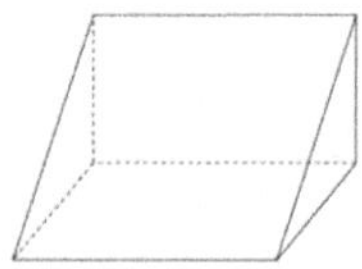

Figure 6.8. Wedge geometry.

Source: Alan, Creative Commons License.

- **Example:** The line of the highest inclination of a flat hillside slopes at an angle of 20° to the horizontal. To reduce the angle of climb, a walker walks on a path on the hillside, which makes an angle of 50° with the line of greatest slope. At what angle to the horizontal does the walker climb on this path? (Figure 6.9).

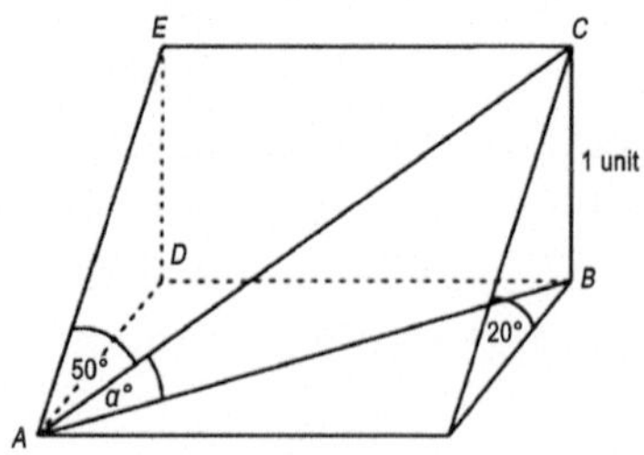

Figure 6.9. Flat hillside slope.

Source: Alan, Creative Commons License.

Let AE represent the line with the highest slope, and AC denotes the trajectory followed by the walking individual. The angle of interest to be determined is denoted as angle BAC. Let the angle be denoted as α°. Given the absence of specified units in the problem, we can assign a unit length to the height BC.

To determine the value of α, identifying an additional length distinct from BC must be identified within triangle BAC. The initial derivation will be obtained from the right-angled triangle DAE, followed by the subsequent derivation from triangle AEC, as seen in Figure 6.10. It should be noted that angle DAE is equal to 20 degrees since line AE exhibits the steepest slope.

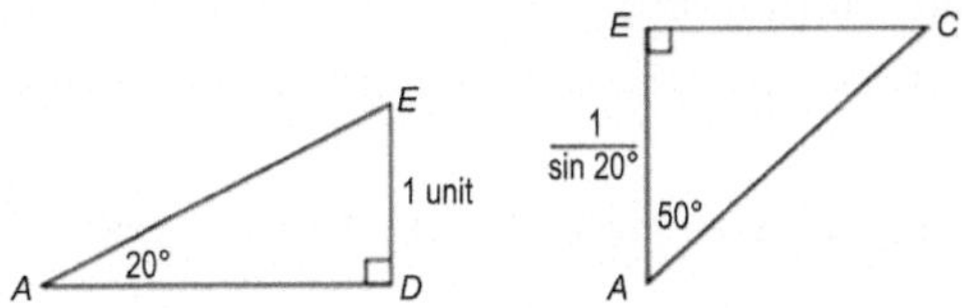

Figure 6.10. Right-angled triangle DAE & triangle AEC.

Source: Alan, Creative Commons License.

In triangle AED

$$\sin 20° = \frac{1}{AE}, so\ AE = \frac{1}{sin20°}$$

In triangle EAC

$$cos50° = \frac{EA}{AC} = \frac{\frac{1}{sin20°}}{AC}, AC = 1/(sin20°cos50°)$$

$$1/AC = (sin20°cos50°)$$

In triangle ABC

$$sina = \frac{1}{AC} = sin20°cos50° = (0.342 \times 0.643), Sina = 0.2199$$

So

$$a° = 12.7$$

The individual engaged in walking moves at an estimated angle of 12.7° relative to the line of maximum incline.

6.6. TRIGONOMETRIC RELATIONSHIPS IN 3D

Like two-dimensional trigonometry, three-dimensional trigonometry analyzes the relationships between angles, sides, and distances in spherical triangles. These relationships allow us to navigate and measure distances in a space that extends beyond flat planes (Desbouvries, 1993; Herranz et al.,2000).

In three dimensions, a triangle formed on the surface of a sphere is known as a spherical triangle. This unique type of triangle deviates from the flat Euclidean triangles we're accustomed to – its sides are arcs along the surface of the sphere, and its angles are measured along these arcs. The angles of a spherical triangle are often denoted by A, B, and C, just like in a flat triangle (Borgomano et al., 2008; Rogers & Pope, 2016).

6.6.1. Law of Cosines for Spherical Triangles

The law of cosines in three dimensions takes on a slightly different form when applied to spherical triangles. For a spherical triangle with sides a, b, and c, and opposite angles A, B, and C, the law of cosines states (Desbouvries, 1993):

$$\cos(a) = \cos(b) \times \cos(c) + \sin(b) \times \sin(c) \times \cos(A)$$

Here, a represents the spherical distance between the endpoints of sides b and c, and A represents the angle opposite side a. This law allows us to relate angles and sides within a spherical triangle, enabling us to solve for unknown angles or distances (Banerjee, 2004).

6.6.2. Law of Sines for Spherical Triangles

The law of sines for spherical triangles establishes a relationship between the sides and angles of a spherical triangle. For sides a, b, and c, and opposite angles A, B, and C, the law of sines can be expressed as:

$$\frac{Sina}{SinA} = \frac{Sinb}{SinAB} = \frac{Sinc}{SinC}$$

This law allows us to determine the ratios of sides and angles within a spherical triangle, enhancing our ability to work with the complexities of spherical geometry (Schwartz, 2010).

6.7. 3D TRIGONOMETRY IN ASTRONOMY AND NAVIGATION

The night sky has fascinated humans for centuries, and astronomy relies on precise measurements and calculations to uncover the universe's mysteries. Three-dimensional (3D) trigonometry has revolutionized both astronomy and navigation, allowing for unprecedented insights into the cosmos while enhancing accuracy and precision in Earth's surface traversal (Maury, 1845; Felski, 2011).

6.7.1. The Parallax Equation

The parallax angle *p*, measured in arcseconds, is half of the angular shift observed. Using the principles of trigonometry, the parallax angle is related to the distance to star *d* and the baseline *B* between Earth's two positions (Rosquist, 1988).

$$p = \text{atan}\left(\frac{d}{B}\right)$$

Here, "atan" represents the arctangent function. The distance to the star can be calculated by rearranging the equation:

$$d = \frac{1}{B}\tan(p)$$

By employing precise measurements of parallax angles and the baseline, astronomers can ascertain the distances to neighboring stars with exceptional precision (Faherty et al., 2007; Bobylev & Bajkova, 2010).

6.7.2. The GPS Equations

By utilizing signals received from a minimum of four satellites, a GPS receiver can employ mathematical equations to ascertain its precise geographical coordinates, including latitude, longitude, and altitude. The signal sent by each satellite encompasses a spherical area that represents the potential places within its range. The point of intersection between these spheres accurately determines the receiver's location (Gabor & Nerem, 1999; Dietrich, 2017).

The distance to each satellite may be estimated by utilizing the pseudo-range (ρ), the speed of light, and the satellite's coordinates (x, y, z).

$$d = c \times \rho$$

The GPS receiver's equations to determine its location depend on the satellite's precise position and the receiver's distance (Schödlbauer.,1993; Borji et al., 2020).

In conclusion, three-dimensional trigonometry has greatly improved the understanding of the cosmos and Earth navigation. 3D trigonometry is essential in many scientific fields, from astronomical distances to GPS satellite locations. With the advancement of technology and understanding, 3D trigonometry will continue to improve precision, facilitate exploration, and illuminate the cosmos.

6.8. ADVANCED 3D TRIGONOMETRY

Advanced 3D trigonometry provides various methods and techniques needed in physics, engineering, computer graphics, and other fields. It uses complicated equations and applications to explain three-dimensional space (Abualnour et al., 2018).

6.8.1. Vectors and Dot Products

One of the cornerstones of advanced 3D trigonometry is the concept of vectors and their dot products. Vectors show direction and magnitude in three-dimensional space, allowing us to portray displacement, forces, velocities, and other phenomena. The degree of alignment between two vectors is measured using the dot product, a basic mathematical procedure. This is especially important when evaluating vector angles.

The mathematical expression for the dot product of two vectors, A and B, is:

$$A \cdot B = |A| \times |B| \times \cos(\theta)$$

The symbol θ represents the angle between vectors. Knowing the vector magnitudes and dot product allows one to calculate the Cosine of an angle using this equation. This notion is commonly used in physics, where understanding forces and their directions are crucial. Lighting and shading computations in computer graphics use it extensively (Donevska-Todorova, 2015).

6.8.2. Cross Products and Determinants

The cross product of two vectors, A and B, yields a new vector that is perpendicular to both input vectors and follows the right-hand rule. This operation finds utility in calculating areas of parallelograms, computing torque, and understanding the interaction of magnetic fields and currents in electromagnetism.

The magnitude of the cross-product is given by:

$$|A \times B| = |A| \times |B| \times \sin(\theta)$$

Here, θ is again the angle between the vectors. The resulting vector points in a direction perpendicular to both A and B and follows a specific orientation determined by the right-hand rule. The cross-product can also be represented using determinants, emphasizing the connection between 3D trigonometry and linear algebra (Ziener et al., 2015).

6.9. APPLICATIONS INVOLVING RIGHT TRIANGLES

6.9.1. Solve Right Triangles

A triangle can be called a right triangle if side "a" represents angle A, side b represents angle B, and side c represents the hypotenuse common, as seen in Figure 6.11. Solving a right triangle requires finding its unknown side lengths and angles. Side dimensions should be rounded to two decimal places and angles to one decimal place per convention. Make sure the calculator is in degree mode. To compute the angles and sides of a right triangle, you must know one of the acute angles, A or B, and a side length or two side lengths. The method uses the Pythagorean Theorem and the fact that the sum of triangle angles equals 180°. Thus, right triangle angles A and B equal 90°.

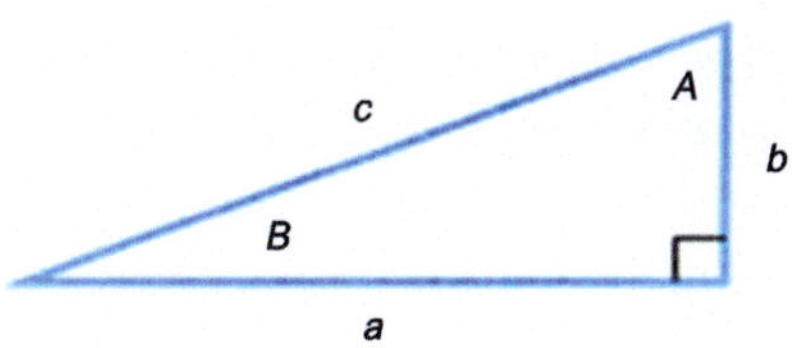

Figure 6.11. Right angle triangle.

Source: Michael, Creative Commons License.

For the right triangle, as shown in the figure we have,

$$c^2 = a^2 + b^2 \; A + B = 90°$$

Example: Use Figure 6.12. If b = 2 and A = 40°, find a, c, and B.

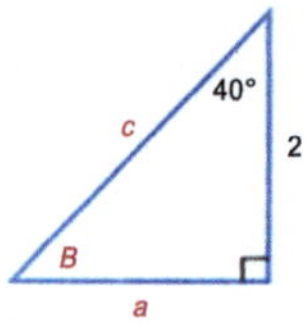

Figure 6.12. Triangle ACB.

Source: Michael, Creative Commons License.

Solution: Since A = 40° and A + B = 90°, we find that B = 50°. To find sides a and c, we use the facts that

$$tan40° = \frac{a}{2} and\; cos40 = \frac{2}{c}$$

Now solve for a and c.

$$\boldsymbol{a} = 2 \approx 1.68 \;\boldsymbol{and\; c} = \frac{2}{\boldsymbol{cos}40°} \approx 2.61$$

6.9.2. Sine and Cosine for any Angle

Right triangle trigonometry was used to calculate the lengths of unknown right triangle sides given an angle and a side length.

Example: Finding the inclination of a mountain trail

A linear path connects the Alpine Hotel, situated at an altitude of 8,000 feet, to a picturesque point at an elevation of 11,100 feet. The route measures a total distance of 14,100 feet. What is the gradient of the trail? In the above diagram, the inquiry pertains to determining the measure of angle B. Based on the information shown in Figure 6.13, it is evident that the length of the side opposite angle B and the hypotenuse length are known. The angle B satisfies the given equation.

$$\sin B = \frac{3100}{14{,}100}$$

Using a calculator,

$$B = sin^{-1}\frac{3100}{14,100} \approx 12.7°$$

The inclination (grade) of the trail is approximately 12 .7°.

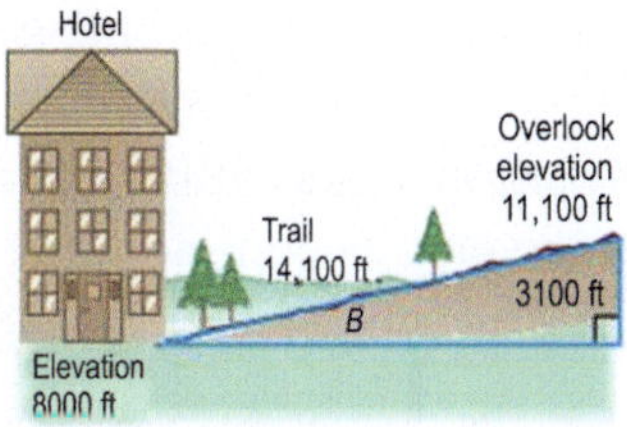

Figure 6.13. Alpine hotel with straight trail leads.

Source: Michael, Creative Commons License.

- **Example:** Finding the bearing of an Airplane
  A Boeing 777 aircraft takes off from O'Hare Airport on runway 2 LEFT, which has a bearing of N20 ° E. After flying for 1 mile, the pilot of the aircraft requests permission to turn 90° and head toward the northwest. The request is granted. After the plane goes 2 miles in this direction, what bearing should the control tower use to locate the aircraft?
- Solution:

Figure 6.14 illustrates the situation. After flying 1 mile from the airport (the control tower), the aircraft is at P. After turning 90° toward the northwest and flying 2 miles, the aircraft is at point Q. In triangle OPQ, the angle e obeys the equation.

$$tan\theta = \frac{2}{1} = 2\, so\, \theta = tan^{-1}2 \approx 63.4°$$

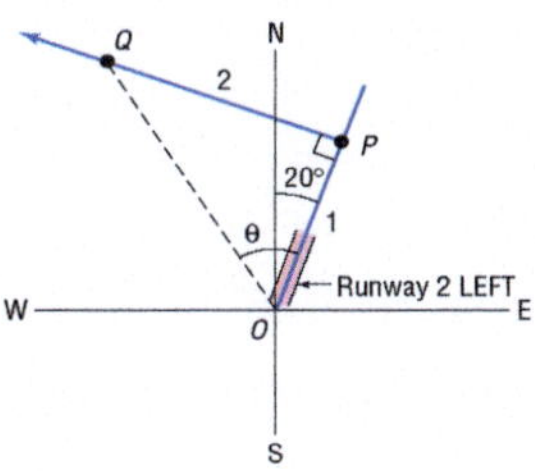

Figure 6.14. Aircraft bearing.

Source: Margaret, Creative Commons License.

The acute angle between the north and the ray OQ is 63.4° – 2 00 = 43.4°. The bearing of the aircraft from 0 to Q is N43.4°W.

6.10. ANGLES OF ANY MAGNITUDE

Angles define the fundamental concepts of geometric connections in trigonometry. Acute, right, and obtuse angles have been previously categorized. This categorization is essential for measuring line segments and plane inclinations. However, this common categorization only scratches the surface of the angles and their roles.

6.10.1. Sine and Cosine for any Angle

In chapter 3, the following construction was given as the basis of the definition of sine and cosine. Draw a circle with a radius of 1 unit and center at the origin O. Draw the radius OP at an angle θ° to the x-axis in an anticlockwise direction (Figure 6.15). Let P have coordinates (x, y).

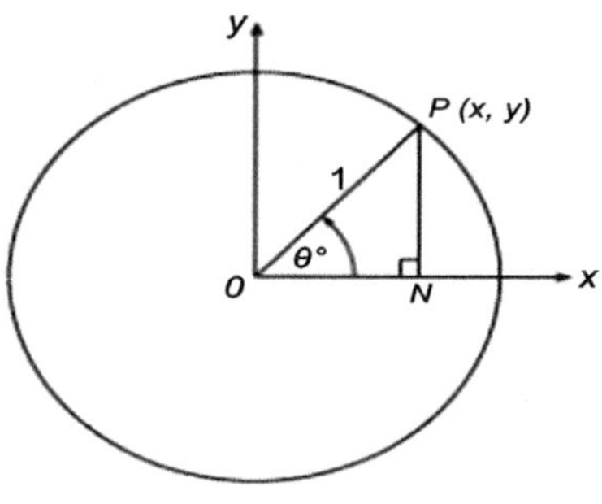

Figure 6.15. Circle with radius 1 unit.

Source: Lial, Creative Commons License.

The equations sin θ° = y and cos θ° = x serve as the formal definitions of the sine and cosine functions, applicable to angles of any magnitude θ°. The arrow denoting the angle θ° in Figure 6.15 highlights the convention of measuring angles in a counterclockwise direction as positive and a clockwise direction as negative. The division of the plane into four quadrants, denoted as 1, 2, 3, and 4, as seen in Figure 6.16, proves to be advantageous.

For angles like 60°, 210°, and -40°, the quadrant in which the radius is placed may be correlated. Figure 6.17 shows Cartesian coordinate angle placement. In particular, it shows that a first quadrant angle of 60° is in quadrant 1. The third quadrant has a 210° angle, and the fourth a -40° angle.

Angles beyond 360° can also be estimated. An angle of 460° is in the second quadrant, whereas -460° is in the third.

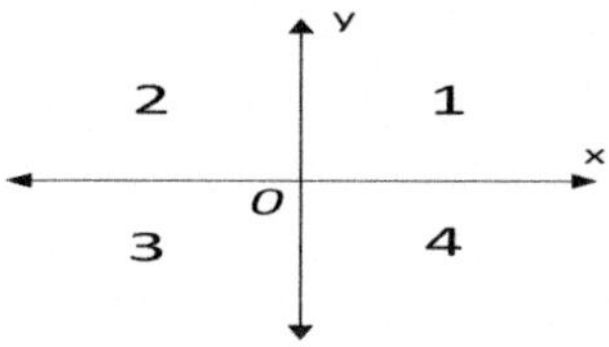

Figure 6.16. Four quadrants.

Source: Lial, Creative Commons License.

Sin θ° = y indicates that when θ° is in the first or second quadrant, the y-coordinate of point P is positive, resulting in sin θ° > 0. However, sin θ° < 0 when θ° is in the third or fourth quadrant. According to the definition of Figure 6.17, the sine of 60° is greater than 0, the sine of 210° is less than 0, and the sine of -40° is less than 0. Calculators make it easy to verify these claims. The definition of Cosine of an angle θ° shows that when 0° is in the first or fourth quadrant, the x-coordinate of point P is positive, resulting in cos θ° > 0. Conversely, cos θ° < 0 for angles in the second or third quadrant. It is clear that the cosine of 60 degrees is larger than zero, 210 degrees is less than zero, and negative 40 degrees is greater than zero. Again, a calculator can validate these computations.

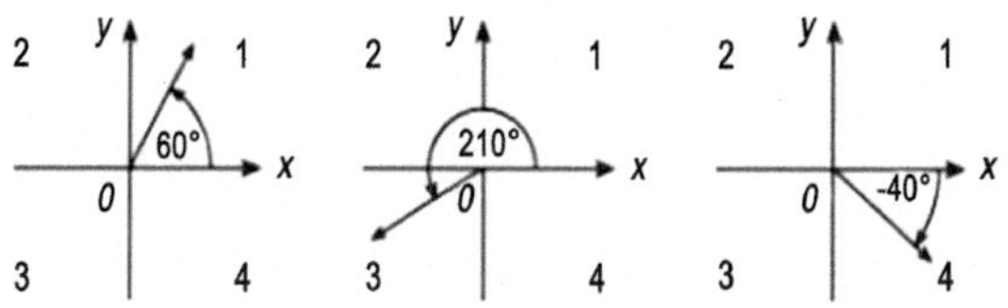

Figure 6.17. Quadrants with 60,210, 40 degrees.

Source: Lial, Creative Commons License.

6.10.2. Sine and Cosine for Multiples of 90°

The easiest way to find the sine and cosine of angles such as 90°, 540° and −90° is to return to the definitions, that is sin θ° = y and cos θ° = x (Figure 6.18). Then you see from the left-hand diagram that the radius for 90° ends at (0, 1), so sin 90° = 1 and cos 90° = 0. Similarly, the radius for 540° ends up at (−1, 0), so sin 540° = 0 and cos 540° = −1.

Finally, the radius for −90° ends up at (0, −1), so sin(−90°) = −1 and cos(−90°) = 0.

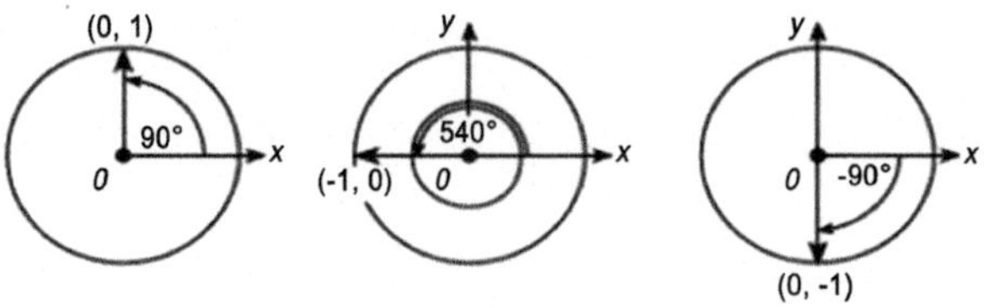

Figure 6.18. Sine and Cosine of angles.

Byjaus, Creative Commons License.

6.10.3. Graphs of Sine and Cosine

As the sine and cosine functions are defined for all angles, you can draw their graphs. The figure 6.19 shows the graph of $y = \sin \theta°$ drawn for values of θ° from −90 to 360. You can see that the graph of $y = \sin \theta°$ has the form of a wave. As it repeats itself every 360°, it is said to be periodic, with a period of 360°. The value of sin θ° is positive for first and second quadrant angles and negative for third and fourth quadrant angles.

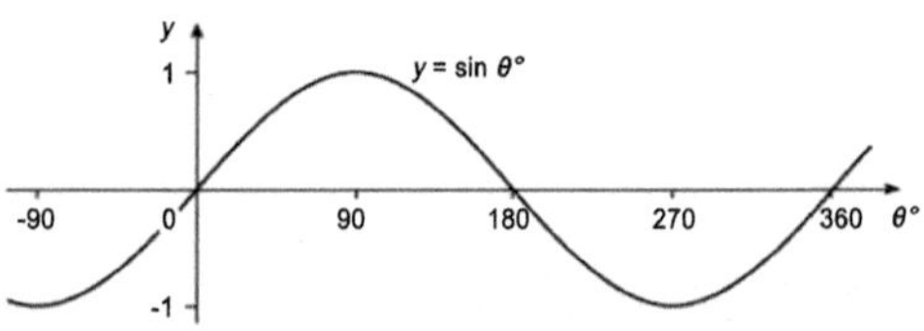

Figure 6.19. Graph of y=sin è °.

Source: Robert, Creative Commons License.

Figure 6.20 shows the graph of $y = \cos \theta°$ drawn for values of θ° from −90 to 360. As you can see, the graph of $y = \cos \theta°$ also has the form of a wave. It is also periodic, with a period of 360°. The value of cos θ° is positive for first and fourth quadrant angles and negative for second and third quadrant angles. It is the wave form of these graphs and their periodic properties that make the sine and cosine so useful in applications. This point is taken further in physics and engineering.

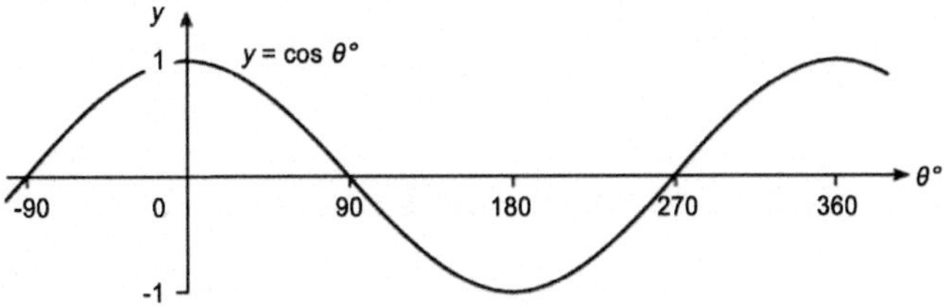

Figure 6.20. Graph of y= cos è°.

Source: Robert, Creative Commons License.

Note that the graph of y = sin θ° shows rotational symmetry concerning the origin. The equation sin θ° = −sin(−θ°) states that the sine of an angle θ° equals the negative sine of its negative (−θ°). For instance, sin45° = −sin(−45°). The graph of y = cos θ° shows y-axis symmetry. The equality cos θ° = cos(−θ°) is valid, as shown by cos45° = cos(−45°). Since the two graphs are translations, distinguishing them might be difficult. This mnemonic technique is meant to help you remember this information: The graph of sin θ°, the sine function, has an extended S shape centered at the origin.

6.10.4. The Tangent of any Angle

The second chapter provided the definition of the tangent for an acute angle. This concept encompasses both positive and negative angles (Figure 6.21).

In the case when the angle θ° is situated in the first quadrant, it may be said that the tangent of θ° is positive. In the context of a second quadrant angle, it can be observed that the value of y is positive while the value of x is negative.

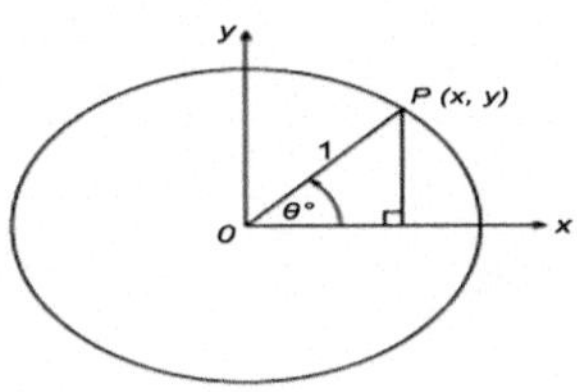

Figure 6.21. Unit circle with è°.

Source: Mark, Creative Commons License.

Consequently, the tangent of the angle θ° is determined to be negative. In the case of a third-quarter angle, where y and x are negative, the tangent of θ° is positive. In the case of a fourth quadrant angle, it is seen that the value of y is negative while x remains positive. Consequently, the tangent of the angle θ° is determined to be negative.

6.10.5. The Graph of the Tangent Function

Similar to the graphical representation of the sine and cosine functions, a graph may also depict the tangent function. It can be shown from Figure 6.22 that, similar to the sine and cosine functions, the tangent function exhibits periodicity. However, it is after 180°, unlike the 360° period of the sine and cosine functions.

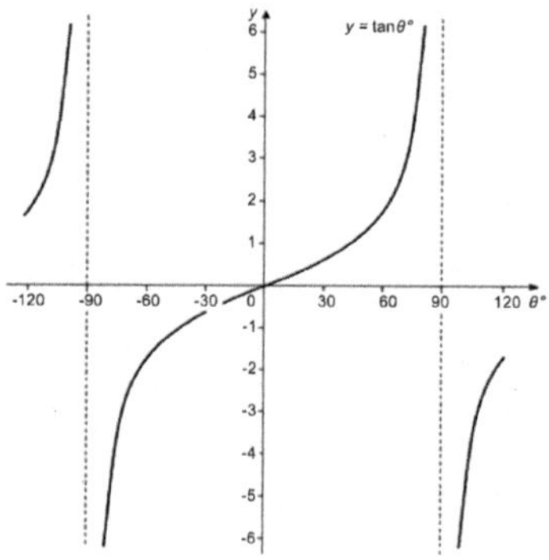

Figure 6.22. Graph of tangent.

Source: Mickel, Creative Commons License.

You can also see that the tangent function is not defined for odd multiples of 90°. There is an important relation between the sine, cosine, and tangent, which you can deduce immediately from their definitions:

$$sin\theta° = y$$

$$cos\theta° = x$$

$$tan\theta° = \frac{y}{x}$$

You can see that,

$$tan\theta° = \frac{sin\theta°}{cos\theta°} = \frac{y}{x}$$

6.11. SUMMARY

This chapter explores spatial perception beyond traditional dimensions. The topic has included historical references, mathematical ideas, and the practical ramifications of extending angles into three dimensions. This chapter covered trigonometry from spherical coordinates through real-world applications.

The chapter introduced spherical coordinates, which help navigate three-dimensional space. Radial distance, polar angle, and azimuthal angle are used in spherical coordinates to define locations on a sphere. This technology improves celestial navigation by providing exact object identification and systematic trajectory mapping across the cosmos.

The main topic was problem-solving, using 3D trigonometry to solve real-world problems. The principles in this chapter allow us to tackle complex spatial connection problems in structure design, navigation over vast terrains, and natural event modeling.

The inquiry became more complicated with trigonometric relationships in three dimensions. The Pythagorean theorem may be used to calculate distances using squared side lengths in three-dimensional space. The law of cosines has been extended to calculate angles and sides in spherical triangles. The links above provide the basis of three-dimensional trigonometry, allowing us to understand the universe's complex geometry. From parallax measurements to GPS-enabled Earth navigation, 3D trigonometry is essential to precision in exploration and comprehension.

Visuals are essential in three-dimensional trigonometry for understanding and connecting abstract mathematical concepts to perceptual awareness. Interactive tools, diagrams, and spherical projections simplify complicated spatial interactions. Breathing life into abstract equations lets us explore the complex world of angles, distances, and shapes.

The chapter examined more complicated topics, demonstrating 3D trigonometry's aesthetic appeal beyond its core foundations. Vectors and dot products show how mathematical structures may express and quantify three-dimensional direction and alignment. Cross products and determinants added complexity and illuminated regions, torques, and magnetic interactions.

REFERENCES

1. Abualnour, M., Houari, M. S. A., Tounsi, A., & Mahmoud, S., (2018). A novel quasi-3D trigonometric plate theory for free vibration analysis of advanced composite plates. *Composite Structures, 184*, 688–697.
2. Agui, T., Sugahara, K., & Nakajima, M., (1988). Three—Dimensional computer animation by trigonometric approximation to aperiodic motion. *Systems and Computers in Japan, 19*(5), 82–89.
3. Allen, T. R., & Kupfer, J. A., (2000). Application of spherical statistics to change vector analysis of Landsat data: Southern Appalachian spruce–fir forests. *Remote Sensing of Environment, 74*(3), 482–493.
4. Anwar, M. S., Choirudin, C., Ningsih, E. F., Dewi, T., & Maseleno, A., (2019). Developing an interactive mathematics multimedia learning based on ispring presenter in increasing students' interest in learning mathematics. *Al-Jabar: Jurnal Pendidikan Matematika, 10*(1), 135–150.
5. Apriandi, D., & Krisdiana, I., (2016). Analisis kesulitan mahasiswa dalam memahami materi integral lipat dua pada koordinat polar mata kuliah kalkulus lanjut. *Al-Jabar: Jurnal Pendidikan Matematika, 7*(2), 123–134.
6. Baidoo, J., & Luneta, K., (2023). Examining the teaching strategies of high-performing mathematics teachers in solving challenging 3-dimensional (3D) trigonometry. *International Journal of Research in Business and Social Science (2147–4478), 12*(2), 429–439.
7. Banerjee, S., (2004). Revisiting spherical trigonometry with orthogonal projectors. *The College Mathematics Journal, 35*(5), 375–381.
8. Bartusiak, M., (2017). *Einstein's Unfinished Symphony: The Story of a Gamble, Two Black Holes, and a New Age of Astronomy* (Vol. 1, No. 1, pp. 1–20). Yale University Press.
9. Bird, J. M., (1917). The mathematics of warfare: How trigonometry and calculus are canned for the gunner's convenience. *The Mathematics Teacher, 10*(1), 35–51.
10. Bobylev, V. V., & Bajkova, A. T., (2010). Galactic parameters from masers with trigonometric parallaxes. *Monthly Notices of the Royal Astronomical Society, 408*(3), 1788–1795.
11. Borgomano, J. R., Fournier, F., Viseur, S., & Rijkels, L., (2008). Stratigraphic well correlations for 3-D static modeling of carbonate reservoirs. *AAPG Bulletin, 92*(6), 789–824.

12. Borji, V., Erfani, H., & Font, V., (2020). A combined application of APOS and OSA to explore undergraduate students' understanding of polar coordinates. *International Journal of Mathematical Education in Science and Technology, 51*(3), 405–423.
13. Buckmire, R., (2003). Investigations of nonstandard, Mickens-type, finite-difference schemes for singular boundary value problems in cylindrical or spherical coordinates. *Numerical Methods for Partial Differential Equations: An International Journal, 19*(3), 380–398.
14. Chen, J. P. J., (2010). The evolution of transformation media in spherical trigonometry in 17th-and 18th-century China, and its relation to "Western learning". *Historia Mathematica, 37*(1), 62–109.
15. Chiddarwar, S. S., & Babu, N. R., (2012). Optimal trajectory planning for industrial robot along a specified path with payload constraint using trigonometric splines. *International Journal of Automation and Control, 6*(1), 39–65.
16. Chin, K. E., (2013). *Making Sense of Mathematics: Supportive and Problematic Conceptions with Special Reference to Trigonometry* (Vol. 13, No. 1, pp. 34–90). University of Warwick
17. Chrzanowski, A., (2023). *Implementation of Trigonometric Height Traversing in Geodetic Levelling of High Precision, 7,* 66–88.
18. Chubey, M., Kouprianov, V., L'vov, V., Tsekmeister, S., Bakholdin, A., Markelov, S., & Levko, G., (2016). Orbital stellar stereoscopic observatory project: Motivation and autonomous navigation in the heliocentric transfer and operational orbits. *Mathematical Modeling and Geometry, 4*(3), 31–49.
19. Defi, S. L., & Qohar, A., (2022). *Development of Learning Media Based on Geogebra Software on the Topic of Trigonometric Ratios* (Vol. 3, No. 2, pp. 34–78). Paper presented at the AIP Conference Proceedings.
20. Desbouvries, F., (1993). *Recursive Least-Squares Lattices and Trigonometry in the Spherical Triangle* (Vol. 67, No. 1, pp. 13–55). Paper presented at the 1993 IEEE International Conference on Acoustics, Speech, and Signal Processing.
21. Dietrich, J. T., (2017). Bathymetric structure-from-motion: Extracting shallow stream bathymetry from multi-view stereo photogrammetry. *Earth Surface Processes and Landforms, 42*(2), 355–364.
22. Donevska-Todorova, A., (2015). Conceptual understanding of dot product of vectors in a dynamic geometry environment. *Electronic*

Journal of Mathematics & Technology, 9(3), 54–78.

23. Dunham, W., (1991). *Journey Through Genius: Great Theorems of Mathematics* (Vol. 8, pp. 78–110). John Wiley & Sons.
24. Emmanouil, E., Wei, G., & Dai, J. S., (2016). Spherical trigonometry constrained kinematics for a dexterous robotic hand with an articulated palm. *Robotica, 34*(12), 2788–2805.
25. Faherty, J., Walter, F. M., & Anderson, J., (2007). *The Trigonometric Parallax of the Neutron Star Geminga* (Vol. 10, No. 1, pp. 45–68. Paper presented at the isolated neutron stars: From the surface to the interior.
26. Fang, W. H., & Wu, M. L., (1997). *An Efficient Unified Systolic Architecture for the Computation of Discrete Trigonometric Transforms* (No. 17, pp. 14–77). Paper presented at the 1997 IEEE International Symposium on Circuits and Systems (ISCAS).
27. Felski, A., (2011). Computation of the azimuth of the great circle in cartesian coordinates. *Annual of Navigation,* (18), 45–53.
28. Gabor, M. J., & Nerem, R. S., (1999). *GPS Carrier Phase Ambiguity Resolution Using Satellite-Satellite Single Differences* (No. 11, pp. 34–98). Paper presented at the Proceedings of the 12th international technical meeting of the satellite division of the institute of navigation (ION GPS 1999).
29. Griffin, J., (2014). *Developing Auxiliary Resource Materials to Support the Engage NY Geometry Curriculum* (Vol. 8, pp. 76–220). Louisiana State University and Agricultural & Mechanical College.
30. Guo, J. Y., Yu, L., Liu, X., Kong, Q. L., & Li, G. W., (2011). Automatic trigonometric leveling system based on GPS and ATR. *Applied Mechanics and Materials, 90*, 2897–2902.
31. He, W., Liu, X., Zhao, L., Li, R., & Liu, F., (2021). Compressed sensing-based simultaneous recovery of magnitude and phase MR images via dual trigonometric sparsity. *IEEE Access, 9*, 38001–38009.
32. Herranz, F. J., Ortega, R., & Santander, M., (2000). Trigonometry of spacetimes: A new self-dual approach to a curvature/signature (in) dependent trigonometry. *Journal of Physics A: Mathematical and General, 33*(24), 4525.
33. Huang, H., Peng, X., Du, W., Ding, S. X., & Zhong, W., (2022). Nitrogen oxides concentration estimation of diesel engines based on a sparse nonstationary trigonometric gaussian process regression with maximizing the composite likelihood. *IEEE Transactions on Industrial*

Electronics, 9(2), 33–95.

34. Kopitnik, Jr. T. A., Kaufman, H. H., Haid, R. W., Marano, G. D., & Nugent, G. R., (1987). Technique of stereotactic biopsy of two cranial targets employing spherical coordinates to define a single trajectory. *Stereotactic and Functional Neurosurgery, 50*(1–6), 188–194.
35. Labrunye, E., & Mallet, J. L., (2004). *Improving Seismic Structural Information Using Trigonometric Interpolation* (Vol. 54, pp. 342–543). Paper presented at the SEG International Exposition and Annual Meeting.
36. Lehmensiek, R., (2012). Using the physical structure of a passive aerodynamic attitude-stabilization system as a multi-frequency antenna on nano-satellites. *IEEE Antennas and Propagation Magazine, 54*(3), 39–49.
37. Levi-Civita, T., (1938). *The Trigonometry of Curvilinear Triangles on a Surface* (Vol. 25, No. 4, pp. 54–98). Rice Institute Pamphlet-Rice University Studies.
38. Liang, C., & Wang, Y. Q., (2020). A quasi-3D trigonometric shear deformation theory for wave propagation analysis of FGM sandwich plates with porosities resting on viscoelastic foundation. *Composite Structures, 247*, 112478.
39. Lorenzo, C. F., & Hartley, T. T., (2016). *The Fractional Trigonometry: With Applications to Fractional Differential Equations and Science* (Vol. 76, pp. 113–234). John Wiley & Sons.
40. Lumbantobing, H., (2018). The effect of interactive computer program on student accomplishment in teaching trigonometry. *International Journal of Sciences: Basic and Applied Research (IJSBAR), 42*(3), 38–48.
41. M'Clelland, W. J., & Preston, T., (1897). *A Treatise on Spherical Trigonometry with Applications to Spherical Geometry and Numerous Examples* (Vol. 13, pp. 23–76). Macmillan.
42. Mahmud, A. A., Tanriverdi, T., & Muhamad, K. A., (2023). Exact traveling wave solutions for (2+ 1)-dimensional Konopelchenko-Dubrovsky equation by using the hyperbolic trigonometric functions methods. *International Journal of Mathematics and Computer in Engineering, 3*(1), 47–85.
43. Maury, M. F., (1845). *A New Theoretical and Practical Treatise on Navigation: In Which the Auxiliary Branches of Mathematics and*

Astronomy, Comprised of Algebra, Geometry. etc., are Treated Of, Finding Time, Latitude and Longitude. EC & J. Biddle, 1, 1-19.

44. McCusker, J., Ma, G., & Caserta, F., (2014). *Can we use a MATLAB Application to Improve Student Performance on Trigonometry of Vector Problem Solving?* Paper presented at the Proceedings of the 2014 Zone 1 Conference of the American Society for Engineering Education.
45. Motz, L., Weaver, J. H., Motz, L., & Weaver, J. H., (1991). The geometry of the circle and trigonometry. *Conquering Mathematics: From Arithmetic to Calculus, 8,* 169–199.
46. North, K. E., (1991). *Analysis of Errors in the Satellite Range Measurements Obtained with the Rockwell-Collins Global Positioning Receiver* (Vol. 17, pp. 86–90). Iowa State University.
47. Ortiz-León, G. N., Loinard, L., Kounkel, M. A., Dzib, S. A., Mioduszewski, A. J., Rodríguez, L. F., & Rivera, J. L., (2017). The gould's belt distances survey (GOBELINS). I. Trigonometric parallax distances and depth of the Ophiuchus complex. *The Astrophysical Journal, 834*(2), 141.
48. Perumaal, S., & Jawahar, N., (2012). Synchronized trigonometric S-curve trajectory for jerk-bounded time-optimal pick and place operation. *International Journal of Robotics and Automation, 27*(4), 385.
49. Petrescu, R. V., Aversa, R., Akash, B., Bucinell, R., Corchado, J., Apicella, A., & Petrescu, F. I., (2017). Inverse kinematics at the anthropomorphic robots, by a trigonometric method. *American Journal of Engineering and Applied Sciences, 10*(2), 394–411.
50. Plofker, K., (2002). *Spherical Trigonometry and the Astronomy of the Medieval Kerala School.* Paper presented at the History of Oriental Astronomy: Proceedings of the Joint Discussion-17 at the 23rd General Assembly of the International Astronomical Union, organized by the Commission 41 (History of Astronomy), held in Kyoto.
51. Prabowo, A., Anggoro, R., Adiyanto, R., & Rahmawati, U., (2018). *Interactive Multimedia-Based Teaching Material for Trigonometry* (Vol. 35, pp. 177–290). Paper presented at the Journal of Physics: Conference Series.
52. Qi, Y., Cui, G., & Miao, J., (2012). *New Trigonometric Series Expansion of the Dirac Function and Application for Infinite Wedge Scattering* (Vol. 3, No. 1, pp. 40–76). Paper presented at the 2012 International

Conference on Microwave and Millimeter Wave Technology (ICMMT).

53. Qiyas, M., & Abdullah, S., (2021). Sine trigonometric spherical fuzzy aggregation operators and their application in decision support system, TOPSIS, VIKOR. *Korean Journal of Mathematics, 29*(1), 137–167.

54. Ratnaningsih, N., Santika, S., Hidayat, E., & Nuraeni, R., (2022). *Development of Interactive Multimedia Using Adobe Flash Based on Problem Solving in Trigonometry Learning* (Vol. 1, No. 1, pp. 1–20). Paper presented at the AIP Conference Proceedings.

55. Richter, W. D., (2007). Generalized spherical and simplicial coordinates. *Journal of Mathematical Analysis and Applications, 336*(2), 1187–1202.

56. Robin-Jouan, Y., (1999). The method of coplanar vertices for astronomical positioning: Present applications and future extensions. *Navigation, 46*(4), 235–248.

57. Rogers, L., & Pope, S., (2016). Working group report: A brief history of trigonometry for mathematics educators. *Proceedings of the British Society for Research into Learning Mathematics, 13*, 103–108.

58. Rosquist, K., (1988). Trigonometric parallaxes of distant objects-what they could tell about the universe. *Astrophysical Journal, Part 1 (ISSN 0004-637X)* (Vol. 331, pp. 648–652). Research Supported by the Naturvetenskapliga Forskningsradet of Sweden.

59. Ross, J. A., Bruce, C. D., & Sibbald, T. M., (2011). Sequencing computer-assisted learning of transformations of trigonometric functions. *Teaching Mathematics and its Applications: An International Journal of the IMA, 30*(3), 120–137.

60. Schermerhorn, B. P., & Thompson, J. R., (2019). Physics students' construction and checking of differential volume elements in an unconventional spherical coordinate system. *Physical Review Physics Education Research, 15*(1), 010112.

61. Schödlbauer, A., Glasmacher, H., Heister, H., Krack, K., & Scherer, B., (1993). *Height Transfer Across the Storebælt (Eastern Channel) Using Geometric Levelling, Trigonometric Heighting and Astronomic Methods in Combination with GPS (GPS Levelling).* Paper presented at the Applications of Geodesy to Engineering: Symposium No. 108, Stuttgart, Germany.

62. Schwartz, R. K., (2010). *Al-Qibla and the New Spherical Trigonometry: The Examples of al-B r n and al-Marr kush* (Vol. 15, pp. 65–98).

Paper presented at the Makalah dipresentasikan di Tenth Maghrebian Colloquium Pada History of Arabic Mathematics (COMHISMA10), Tunis.

63. Suárez, M. D., Pavlov, L., & Zelasco, J. F., (2022). Efficiency comparison of quaternions and spherical trigonometry formulas for 3D rotations. *International Journal of Applied and Computational Mathematics, 8*(2), 85.
64. Thompson, K. A., (2007). *Students' Understanding of Trigonometry Enhanced Through the Use of a Real World Problem: Improving the Instructional Sequence* (Vol. 7, pp. 76–90). Illinois State University.
65. Van, B. G., (2012). *Heavenly Mathematics: The Forgotten Art of Spherical Trigonometry* (Vol. 4, pp. 45–89). Princeton University Press.
66. Volder, J. E., (1959). The CORDIC trigonometric computing technique. *IRE Transactions on Electronic Computers,* (3), 330–334.
67. Wu, H., (2013). Teaching geometry in grade 8 and high school according to the common core standards. *Retrieved April, 8*, 2017.
68. Zengin, Y., Furkan, H., & Kutluca, T., (2012). The effect of dynamic mathematics software GeoGebra on student achievement in teaching of trigonometry. *Procedia-Social and Behavioral Sciences, 31*, 183–187.
69. Zenkour, A., (2007). Benchmark trigonometric and 3-D elasticity solutions for an exponentially graded thick rectangular plate. *Archive of Applied Mechanics, 77*(4), 197–214.
70. Ziener, C. H., Kurz, F. T., Buschle, L. R., & Kampf, T., (2015). Orthogonality, Lommel integrals and cross product zeros of linear combinations of Bessel functions. *SpringerPlus, 4*, 1–26.
71. Zucker, C., Schlafly, E. F., Speagle, J. S., Green, G. M., Portillo, S. K., Finkbeiner, D. P., & Goodman, A. A., (2018). Mapping distances across the Perseus molecular cloud using CO observations, stellar photometry, and Gaia DR2 parallax measurements. *The Astrophysical Journal, 869*(1), 83.

CHAPTER 7

Trigonometric Functions and Graphs

Contents

7.1. INTRODUCTION

The present chapter provides a thorough examination of trigonometric equations and their graphical depictions, hence establishing a solid basis for comprehending this fundamental area of mathematics.

The sine, cosine, tangent, cotangent, secant, and cosecant functions are essential mathematical tools used to illustrate different phenomena in terms of graphs (F. R. Moore, 1978). Originating from the discipline of math, these functions have undergone significant development and have become essential tools in several domains, including engineering and physics (Kendal & Stacey, 1996).

The investigation starts with an examination of these functions in the context of angles, explaining their intrinsic characteristics and meanings. Subsequently, we explore the process of transitioning from acute to obtuse angles, revealing the unique characteristics shown by these functions in different situations.

The graphical representations of trigonometry serve as a means to unveil its fundamental nature. The sinusoidal shapes of the cosine and sine functions demonstrate their periodicity and symmetrical properties (Maulyda, Hidayanto, & Rahardjo, 2019). In the context of mathematical analysis, it is noteworthy that the graphs of tangent and cotangent have a dynamic nature, characterized by the presence of asymptotes (Curri, 2012).

7.2. TRIGONOMETRIC FUNCTIONS OF AN ANGLE

To define trigonometric functions for any angle A, the angle is placed in a rectangular coordinate system with angle A at the vertex between r and x. Where, r (positive) is the distance from V to any point Q on the terminal side of A, and (x, y) are the rectangular coordinates of Q.

The six functions of A are then defined by six ratios exactly as in the earlier case for the triangle given in the introduction. Because division by zero is not allowed, the tangent and secant are not defined for angles the terminal side of which falls on the y-axis, and the cotangent and cosecant are undefined for angles the terminal side of which falls on the x-axis. When the Pythagorean equality $x^2 + y^2 = r^2$ is divided in turn by r^2, x^2, and y^2, the three squared relations relating cosine and sine, tangent and secant, cotangent and cosecant are obtained (Ungar, 2001; Calzada & Scariano, 2006) (Figure 7.1) (Table 7.1).

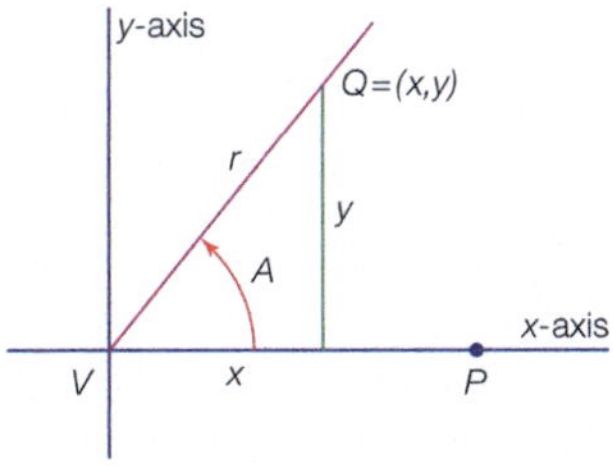

Figure 7.1. Angle in standard position.

Source: Britannica, Creative Commons License.

Table 7.1. A Table for Negative Angles

Input	**Output**
Sin (-A)	- sin A
Cos (-A)	cos A
Tan (-A)	-tan A
Csc (-A)	-csc A
Sec (-A)	sec A
Cot (-A)	-Cot A

Source: Eli Maor, Creative Commons License.

To establish the trigonometric equations for an arbitrary angle *A*, it is necessary to consider the angle within the range of 0 to 2π. In trigonometry, when we examine an angle *A* in its standard position, and there's a point *Q* situated on the opposite side of this angle with coordinates (x, y), an interesting relationship becomes apparent. When we relocate point *Q* to the terminal side of the angle *-A*, which is also in the standard position, its coordinates undergo a specific transformation. In this new position, the coordinates of point *Q* become (x, -y). This geometric observation highlights the symmetry between the two angles *A* and *-A,* in terms of their impact on the y-coordinate of a point on the opposite side. The Aforementioned relationships can also be expressed by stating that cosine and secant functions exhibit even symmetry (symmetry around the y-axis), whereas the remaining four functions demonstrate odd symmetry (symmetry about the origin) (Kamber & Takaci, 2018).

A function derived from trigonometry exhibits uniformity in its value across all coterminal angles. When n represents an integer, it follows that sin (A ± 360n) is equal to sin A. Similar relationships may be observed for the remaining five functions. The findings may be articulated as follows: the trigonometric functions exhibit periodicity, with a period of either 360° or 180° (Agarwal, 2020).

When point Q is located on the opposite side of angle A in standard position, its coordinates (x, y) are transformed to (-y, x) and (y, -x) on the terminal sides of angles $A + 90$ and $A - 90$ in normal position, respectively. As a result, six mathematical formulae establish an equivalence between the value of the complements of set A and the equivalent cofunction of set A.

A line, denoted as line AB, is formed by connecting two unique locations, A and B. The segment of line AB that originates from point A and extends beyond point B is referred to as ray AB. Point A serves as the terminal point of the beam (Figure 7.2).

The two opposing rays (or segments) serve as the sides of the resulting angle, while the shared endpoint represents the center of the angle. The measure of an angle is determined by a rotation around its vertex (Figure 7.3). The determination of this measure involves the rotation of a ray originating from the opposite end of the angle, referred to as the beginning side, to the location of the opposing side, known as the terminal side. A rotation in the counterclockwise direction is associated with a positive metric, whereas a rotation in the clockwise direction is associated with a negative value. The rotational motion has the potential to encompass many full revolutions (Cetin, 2015; Chin & Tall, 2012).

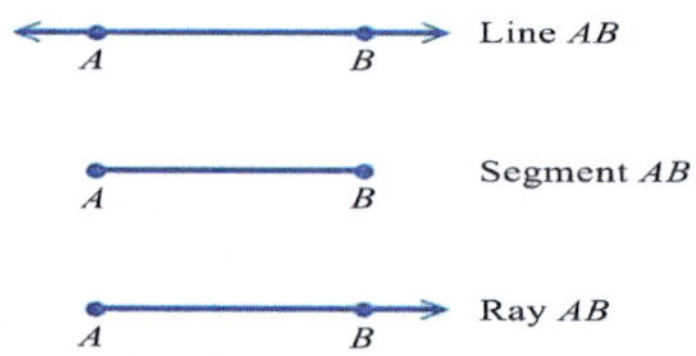

Figure 7.2. Line, segment, and ray AB.

Source: David, Creative Commons License.

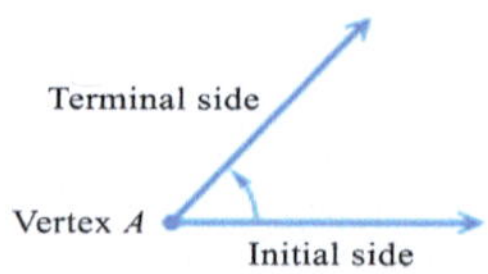

Figure 7.3. Angle generated by rotation about the vertex.

Source: David, Creative Commons License.

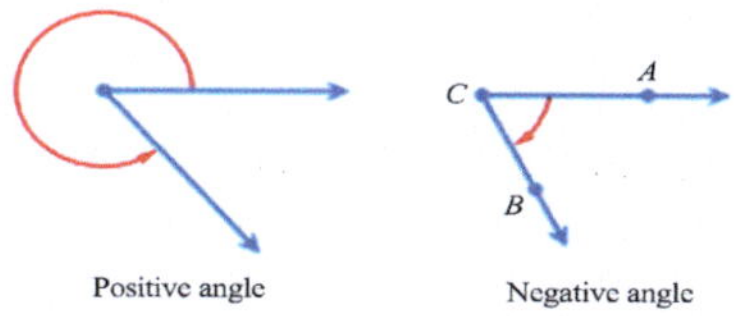

Figure 7.4. Two angles, positive and negative.

Source: David, Creative Commons License.

An angle can be named by using the name of its vertex. For example, the angle on the right in Figure 7.4 can be named angle C. Alternatively, an angle can be named using three letters, with the vertex letter in the middle. Thus, the angle on the right also could be named angle ACB or angle BCA.

7.2.1. Degree Measure

Degree is the common unit of measurement for angles. In the context of degree measurement, the entire turning radius of a ray is normally assigned a value of 360 degrees (Kalanov, 2014; Moore, 2012). In Figure 7.5, it can be observed that the angle becomes the same after completing a full revolution.

$$One\,degree\left(written\,as\,1°\right)represents\,\frac{1}{360}\,of\,a\,rotation.$$

Hence, it can be observed that an angle of 90° corresponds to one-fourth (1/4) of a whole revolution, whereas an angle of 180° corresponds to one-half (1/2) of an entire revolution. An acute angle is defined as an angle that measures between 0° and 90°. A right angle is defined as an angle that measures precisely 90°. The sign "m" is frequently employed to represent the measurement of 90° at the northernmost point of a right angle.

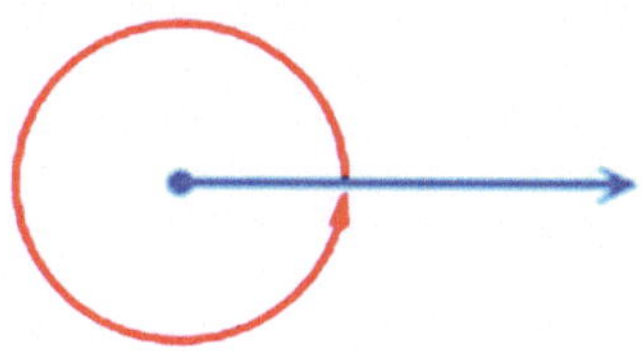

Figure 7.5. The angular measure of a full revolution of a beam is 360°. An angle of 1° is obtained by dividing a full revolution into 360 equal parts.

Source: Margaret, Creative Commons License.

An angle that has a measure higher than 90° but does not exceed 180° is classified as an obtuse angle, whereas an angle with a measure of approximately 180° is referred to as a straight angle (Maknun, Rosjanuardi, & Jupri, 2019).

In Figure 7.6, we use the Greek letter θ (theta) to name each angle.

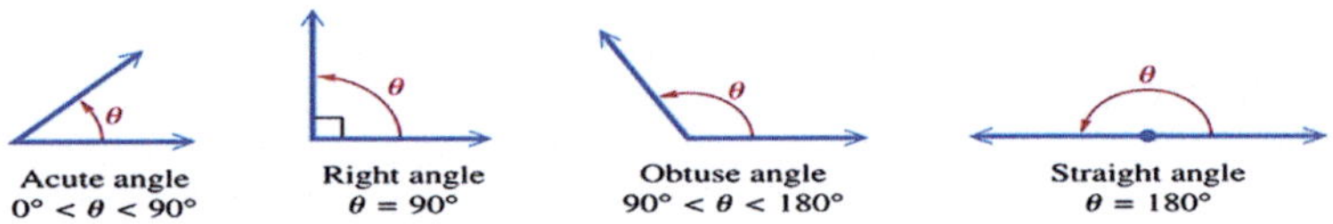

Figure 7.6. Different values of angle (θ).

Source: Lial, Creative Commons License.

If the sum of the measures of two positive angles is 90°, the angles are complementary, and the angles are complements of each other. Two positive angles with measures whose sum is 180° are called supplementary angles (Bressoud, 2010; Moore, 2010).

7.2.2. Standard Position

An angle is said to be in standard position when its vertex is located at the origin of a coordinate plane, and its starting side is positioned along the positive x-axis (Wallis, 1974; Walsh et al., 2017). The angles depicted in Figure 7.7 are situated in the standard position. An angle in a typical position is commonly referred to as lying in the quadrant that corresponds to the location of its terminal side. An acute angle is situated inside quadrant I, as depicted in Figure 7.7(a), whereas an obtuse angle is positioned within quadrant II, as illustrated in Figure 7.7(b). In Figure 7.7, the shown ranges of angle measurements for each quadrant are presented within the interval of 0°<θ<360°.

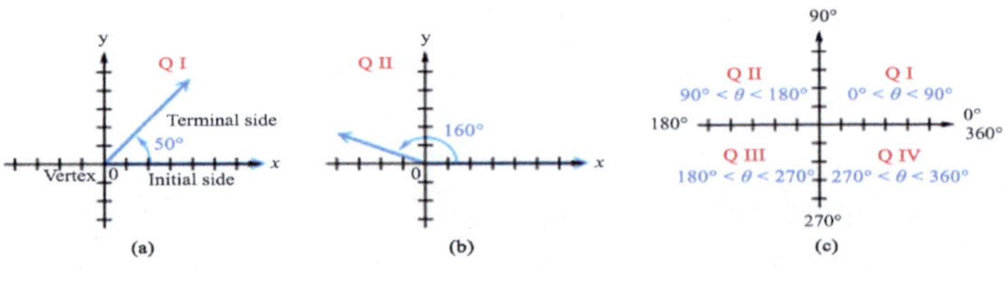

Figure 7.7. Measurement of angles in different quadrants.

Source: Schneider, Creative Commons License.

7.2.3. Coterminal Angles

The whole revolution of a beam corresponds to an angular measurement of 360°. By persisting with the process of rotation, it is possible to generate angles that exceed a measurement of 360° (Stitz & Zeager, 2011). The angles depicted in Figure 7.8, measuring 60° and 420°, have a shared beginning side and terminal side; however, they differ in terms of the extent of rotation. These angles might be referred to as coterminal angles (King et al., 2020). According to the illustration presented in Figure 7.8, it can be observed that angles measuring 110° along with 830° are conterminal.

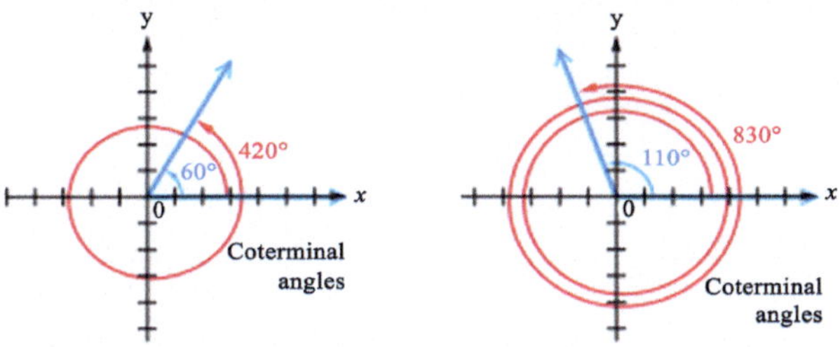

Figure 7.8. (a) Angle with measures of 60 ° & 420° , (b) angles with measures 110 ° & 830 ° .

Source: Daniels, Creative Commons License.

7.2.4. Tables of Natural Functions

The values of trigonometric functions must be easily accessible for any specified angle for their application in actual scenarios. Multiple trigonometric identities demonstrate that the significance of the functions of trigonometry for all angles can be easily determined based on the values for angles ranging from 0° to 90°. Hence, it is suitable to compile a tabular representation encompassing the corresponding digits of the cosine, sine, and tangent functions for angles ranging from 0° to 45°, namely those that are whole multiples of a suitable unit (Moritz, 1913; Fogiel, 1984).

In cases when angles are not exactly multiples of the unit, it is possible to estimate the actual values of function values by interpolation. Due to the typically irrational nature of function values, they are commonly represented in decimal form inside tables, rounded to a suitable degree of precision. In many applications, it is often acceptable to utilize numerical values with up to five decimal places, as table 7.2 providing data with this level of precision are widely available. Merely relying on basic geometric principles is adequate for establishing the appropriate values of the functions of trigonometry for angles that include angular values of 0°, 30°, 45°, 60°, and 90°. The values representing the tangent, cosine, and sine functions are presented in a tabular format (Bond, 1921; Garibaldi, 2008).

Table 7.2. Common Angles for Trigonometry Functions

Functions–Angles	0°	30°	45°	60°	90°
Sin	0	½	$\frac{\sqrt{2}}{2}$	$\frac{\sqrt{3}}{2}$	1
Cos	1	$\frac{\sqrt{3}}{2}$	$\frac{\sqrt{2}}{2}$	$\frac{1}{2}$	0
Tan	0	$\frac{\sqrt{3}}{3}$	1	$\sqrt{3}$	Undefined

Source: Eli Maor, Creative Commons License.

7.3. ANGLE RELATIONSHIPS AND SIMILAR TRIANGLES

In the diagram presented below, both sides of the circle RMQ have been stretched to create an additional angle denoted as RMQ. The angles NMP and RMQ are considered to be vertical angles. An additional set of vertical angles, namely NMQ and PMR, are also created. Vertical angles possess the following properties (Price & van Jaarsveld, 2017).

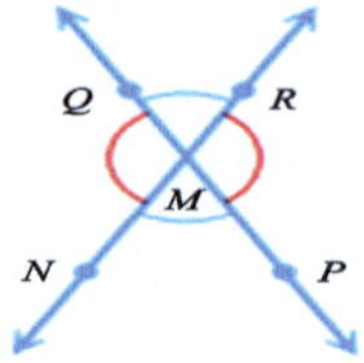

Figure 7.9. Vertical angles.

Source: Michael, Creative Commons License.

Parallel lines are a pair of lines that are coplanar but do not intersect. The diagram in Figure 7.10 illustrates the presence of parallel lines denoted as *m* and *n*. When a line, denoted as *q*, overlaps two parallel lines, it is referred to as a transversal line. Figure 7.10 depicts a transversal line intersecting a pair of parallel lines, resulting in the formation of eight angles. These angles are conveniently labeled with corresponding numbers (Adamek et al., 2005; Liu et al., 2011).

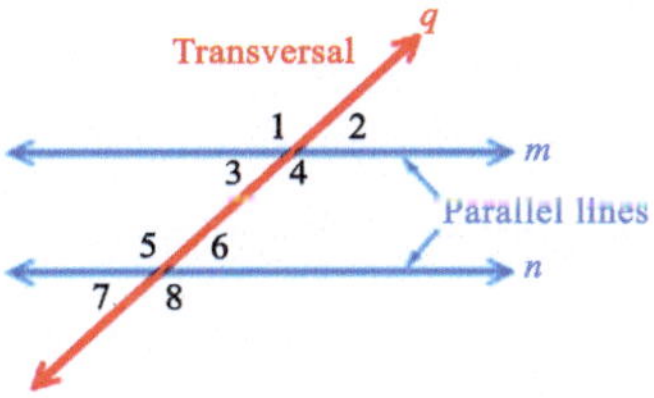

Figure 7.10. Parallel & transversal lines.

Source: Lial, Creative Commons License.

Let's explore a compelling rationale supporting the idea of similar angles and triangles by studying a triangle of arbitrary dimensions obtained through the process of cutting a sheet of paper (May & Courtney, 2016; Kamber & Takaci, 2018). The recommended action in Figure 7.11 involves the removal of each of the corners from the triangle. The ability to rearrange the components in such a manner that all three positions align to create a straight angle, measuring 180°, is demonstrated in Figure 7.11 (Cavey & Berenson, 2005; Challenger, 2009).

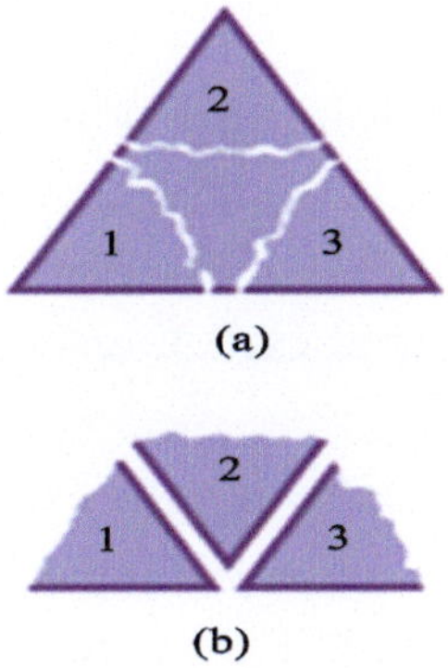

Figure 7.11. Each corner is torn from a triangle.

Source: Michael, Creative Commons License.

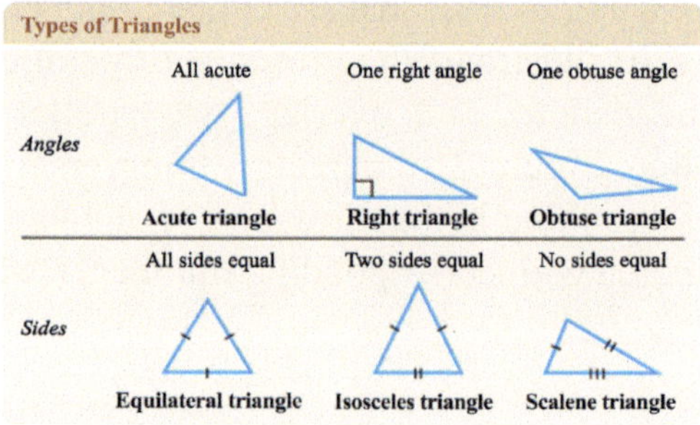

Figure 7.12. Types of triangles.

Source: Raymond, Creative Commons License.

7.4. TRIGONOMETRIC FUNCTIONS

To establish an understanding of the six functions of trigonometry, we start the process by considering an angle *u* positioned in standard form and proceed to select a point P on the opposite side of angle *u*, which possesses coordinates (x, y) (Weber, 2008; Bornstein, 2020). Point *P* should not be the center of the angle (Figure 7.13). The x-axis intersects point *Q*, forming a right triangle with a vertex at *O, P,* and *Q*. The distance, denoted as *r*, between the point *P(x, y)* and the origin *O(0, 0)* may be determined by employing the distance formula.

$$r = \sqrt{(x-0)^2 + (y-0)^2}$$

$$r = \sqrt{x^2 + y^2}$$

It should be noted that the value of r is more than zero, as it represents the undirectional distance.

The six trigonometric functions associated with angle *u* are sin(*u*), cos(*u*), tan(*u*), cot(*u*), sec(*u*), and csc(*u*), respectively (Kamber & Takaci, 2018).

- Example: Finding Function Values of an Angle:

  The end on the opposite side of angle *u*, when in standard position, intersects with the point (8, 15). Determine the values of the eight trigonometric coefficients (sine, cosine, tangent, cosecant, secant, and cotangent) for a given angle θ.
- **Solution:** The diagram depicted below illustrates the angle θ and

the triangular shape that is created when a perpendicular line is extended through the intersection of two points (8, 15) to intersect the x-axis. The coordinates of (8, 15) indicate that it is situated eight degrees to the left of the y-axis center and 15 units beneath the x-axis. Utilizing $r = \sqrt{x^2 + y^2}$

$$r = \sqrt{8^2 + 15^2}$$

$$= \sqrt{64 + 225}$$

$$= \sqrt{289} = 17$$

The values of the six trigonometry functions of angle θ can be determined as follows:

$$sin\theta = \frac{y}{r} = \frac{15}{17} os\theta = x / r = 8/17$$

$$tan\theta = \frac{y}{x} = \frac{15}{8} sc\theta = r / y = 17/15$$

$$sec\theta = \frac{r}{x} = \frac{17}{8} ot\theta = x / y = 8/15$$

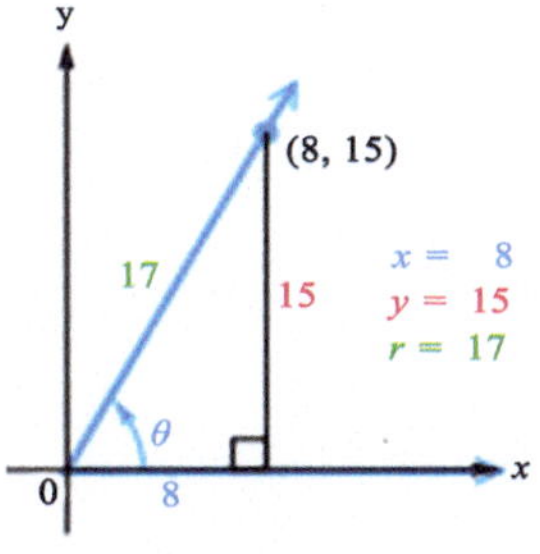

Figure 7.13. Triangle formation using a point and an angle.

Source: Walter, Creative Commons License.

- **Quadrant Angles:** When the terminal end of an angle in the normal position is aligned with the y-axis, all points on that terminal side will have an x-coordinate of 0 (Sanna et al., 2011). Likewise, it can be shown that an angle whose terminal side lies on the x-axis would consistently possess a y-coordinate of 0 at any given position along the terminal side. The trigonometric functions involve the values of x and y in their denominators, and according to the mathematical convention, a fraction is considered meaningless if its denominator is equal to zero. As a result, some trigonometry function values for quadrantal angles, which are angles with their terminal side lying on an axis, are undefined (Mokhtar et al., 2018). When calculating the values of trigonometric functions for quadrantal angles, the diagram shown below can be utilized to get the corresponding ratios. Since any point located on the side of the terminal is applicable, it is advantageous to select a point situated one unit away from the axis, with a radial distance of 1 (Koks, 2022). To find the values of the function of a quadrantal position, it is necessary to find the position on the terminal side. Subsequently, one must select one of the four points that lie on said terminal side and subsequently employ the definitions that include x, y, and r (Zainal & Ismail, 2010) (Figure 7.14).

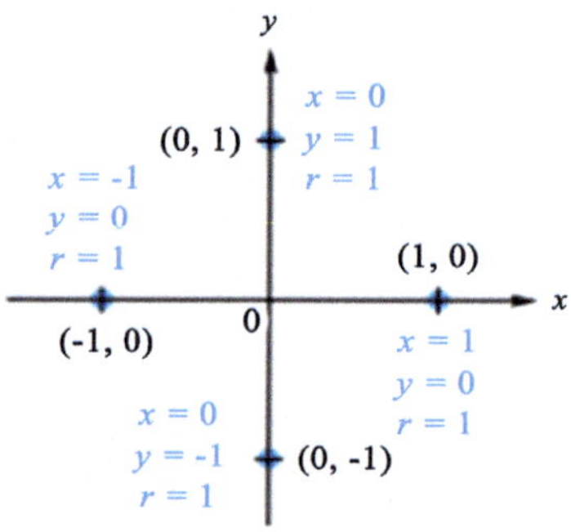

Figure 7.14. Quadrant angles.

Source: Gloria, Creative Commons License.

7.4.1. Trigonometric Function of Acute Angles

7.4.1.1. Right Angled Triangles

The trigonometric functions are defined using angles in standard position. An alternative method can be employed in dealing with them (Ali et al., 2016). Right triangles can be represented by proportions of the widths of their sides.

The diagram presented below depicts a right angle *(A)* in the standard position. The determination of the trigonometric function corresponding to angle *A* necessitates the consideration of the variables *x, y,* and *r*. As seen in Figure 7.15, the variables *x* and *y* represent the measurements of the two legs for the right triangle ABC, while the variable *r* denotes the measurement of the hypotenuse. The side opposing angle *A* is denoted as *y*, whereas the side next to angle *A* is denoted as *x*. The lengths of the sides are employed to substitute the variables *x* and *y* in the trigonometric function definitions, while the hypotenuse is used to replace the variable *r*, resulting in the subsequent definitions based on right triangles (Raychev, 2015; Ali et al., 2016) (Figure 7.16).

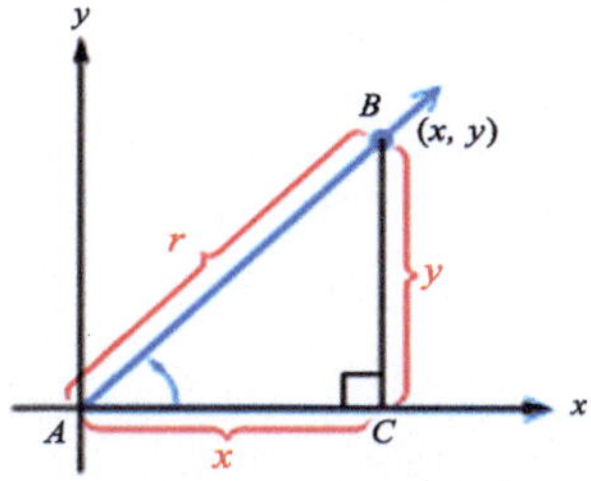

Figure 7.15. Acute angle in standard position.

Source: Sullivan, Creative Commons License.

Let A represent any acute angle in standard position.

$$\sin A = \frac{y}{r} = \frac{\textit{side opposite } A}{\textit{hypotenuse}} \qquad \csc A = \frac{r}{y} = \frac{\textit{hypotenuse}}{\textit{side opposite } A}$$

$$\cos A = \frac{x}{r} = \frac{\textit{side adjacent } A}{\textit{hypotenuse}} \qquad \sec A = \frac{r}{x} = \frac{\textit{hypotenuse}}{\textit{side adjacent } A}$$

$$\tan A = \frac{y}{x} = \frac{\textit{side opposite } A}{\textit{side adjacent } A} \qquad \cot A = \frac{x}{y} = \frac{\textit{side adjacent } A}{\textit{side opposite } A}$$

Figure 7.16. Trigonometric ratios for a right-angled triangle.

Source: Sullivan, Creative Commons License.

7.4.2. Trigonometric Function of Non-Acute Angle

Each non-quadrantal angle in the standard position is accompanied by a positive acute angle known as its reference angle. The point of reference for a given angle, denoted as *u*, is the acute angle formed by the point on either side of degree u and the x-axis, with a positive orientation (Arts et al., 2010;

Liu et al., 2021). Figure 7.17 illustrates many angles, denoted as *u*, situated in regions II, III, and IV, respectively, each spanning less than a complete counterclockwise revolution. Additionally, the reference angle *u* is depicted. In the first quadrant, the angles θ and θ' are equal (Yao & Yang, 2010). If an angle *u* possesses a negative value or has a measure beyond 360°, and its equivalent angle may be determined by initially identifying its mutually exclusive angle that lies within the range of 0° to 360°. Subsequently, the diagrams presented in Figure 7.17 can be utilized for further analysis.

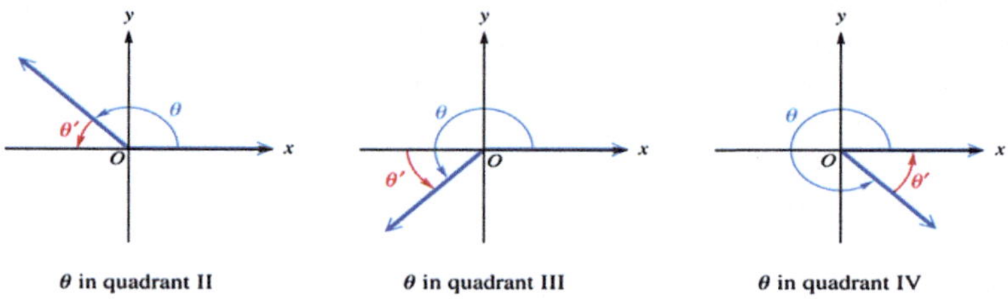

Figure 7.17. Angle è in each quadrant.

Source: Raymond, Creative Commons License.

- **Example:** Determine the reference angle for each given angle: (a) 218° (b) 1387°
- **Solution**: Solution As seen in Figure 7.18, the acute angle formed between the terminating side of this inclination and the axis x is positive.

$218° - 180° = 38°$

For $\theta = 218°$ the reference angle $\theta' = 38°$

a. Initially, it is necessary to determine a coterminal angle within the range of 0° to 360°. To determine the quotient of 1387° divided by 360°, the result is around 3.9. To begin the process, it is necessary to operate subtracting 360° on three separate occasions, as dictated by the integer value of 3 in the decimal number 3.9.

$$1387° - 3\times360° = 1387° - 1080° = 307°$$

The orientation angle for 307° (and therefore for 1387°) is 360 – 307 = 53.

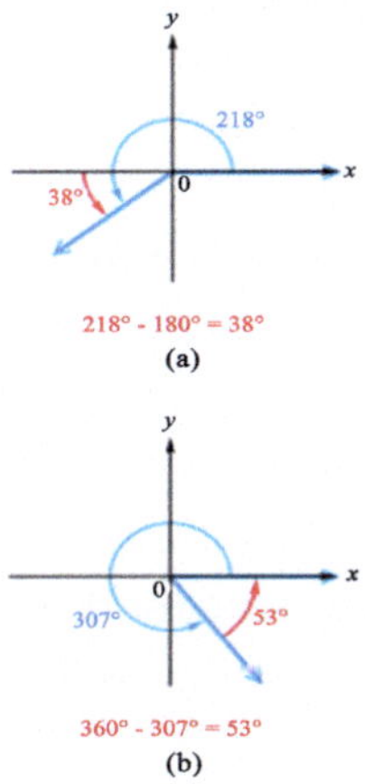

Figure 7.18. (a) Positive acute angle, (b) reference angle for 307 degrees.

Source: Raymond, Creative Commons License.

7.5. GRAPHS OF SINE AND COSINE FUNCTION

7.5.1. Periodic Functions

Numerous phenomena encountered in everyday life have an evident recurrence characterized by a regular sequence, including weather patterns, tidal movements, and the duration of daylight. The periodic nature of the cosine and sine functions arises from their ability to cycle through their values over time in a consistent fashion (Martínez-Planell & Delgado, 2016; Yang & Sianturi, 2017). The graph depicted in Figure 7.19 illustrates a periodic pattern that accurately mimics the rhythmic nature of a typical pulse.

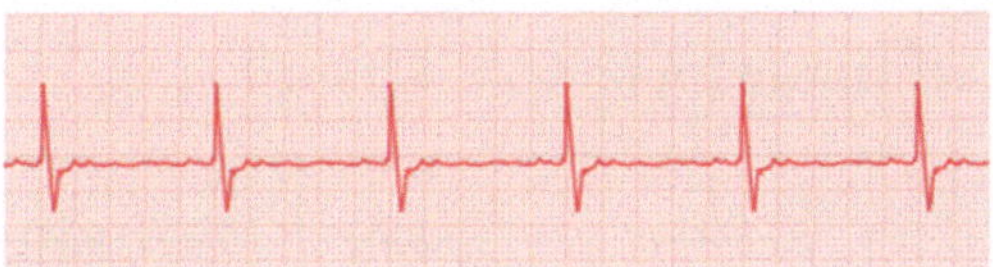

Figure 7.19. Normal ECG.

Source: Margaret, Creative Commons License.

A periodic function is represented as:

$$f(x) = f(x + np),$$

Where x is a real number inside the domain of the function f;

n is any integer;

p is a significant real number.

The time dimension of the function is determined by the smallest positive value of p (Martínez-Planell & Delgado, 2016).

The total angular value of a single circle is equal to 2π. Therefore, the minimum value of π that occurs when both cosine and sine functions exhibit repetition is 2π. Hence, it can be concluded that both cosine and sine functions exhibit periodicity with a period of 2π. Furthermore, it is valid to assert that the subsequent propositions hold for any number n.

7.5.2. Graph of the Sine Function

It is well-established that, for every real integer *s*, the arbitrary points on the circle can be determined using sine and cosine functions (e.g., cos(s), sin(s)) (DeJarnette, 2018; Maulyda et al., 2019). Refer to the diagram presented in the figure 7.20 to see the coordinates of different points.

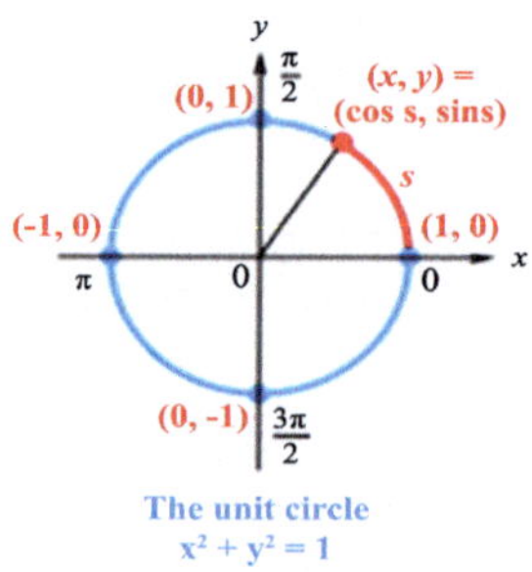

Figure 7.20. Unit circle.

Source: Margaret, Creative Commons License.

Table 7.3. Results are Drawn From the Unit Circle

As s Increases From	Sin(s)	Cos(s)
$0\ to\ \frac{\pi}{2}$	Increases from 0 to 1	Decreases from 1 to 0
$\frac{\pi}{2}\ to\ \pi$	Decreases from 1 to 0	Decreases from 0 to -1

$\pi\ to \dfrac{3\pi}{2}$	Decreases from 0 to -1	Increases from -1 to 0
$\dfrac{3\pi}{2} to\ 2\pi$	Increases from -1 to 0	Increases from 0 to 1

Source: Lial, Creative Commons License.

To mitigate any ambiguity during the graphical representation of the sine function, the variable *x* is employed instead of *s*, aligning with the conventional usage of letters within the xy-coordinate system. The process of identifying particular amounts of *x* and determining the related value of sin(x) results in the creation of a table, as seen in Figure 7.21.

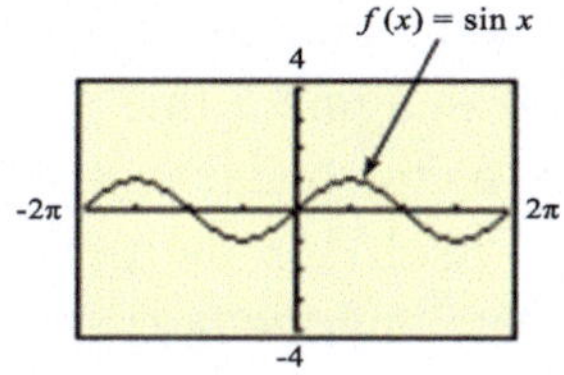

Figure 7.21. Value of sin X.

Source: Sullivan, Creative Commons License.

To generate the conventional graph, as seen in Figure 7.21, it is necessary to plot the data points provided in the table. To do this, apply the principle of symmetry and connect them with a continuous curve. The function y = sin(x) exhibits periodicity with a period of 2π and possesses a domain of (-∞, ∞). Consequently, the graph of this function maintains a consistent pattern in both positive and negative directions. The graphical representation depicted herein is sometimes referred to as a wave that is sine or sinusoid (Demir & Heck, 2013).

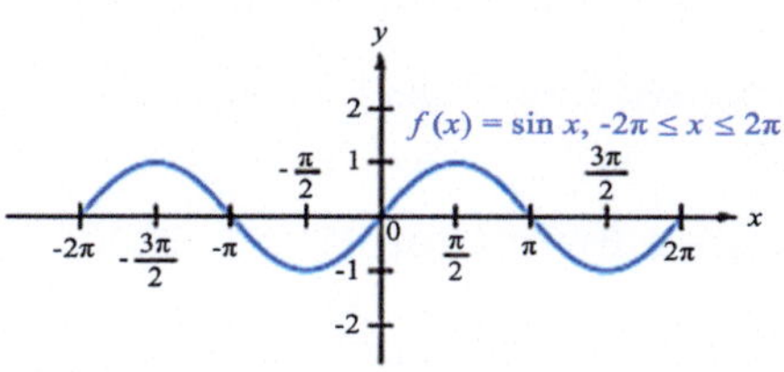

Figure 7.22. Sine graph.

Source: Rob Hainer, Creative Commons License.

Here are the characteristics of the graph (Figure 7.22):

- The graph exhibits continuity over its complete domain, which extends from negative infinity to positive infinity.
- The x-intercepts of the function may be expressed as multiples of p, where p is a constant and n is an integer.
- The period of the given phenomenon is 2π.
- The graph exhibits symmetry about the origin, indicating that the function is odd. For every x inside the specified domain, the equation sin(-x) is equivalent to –sin(x) (Fogiel et al. 1984).

The sine function has a close relationship with the unit circle. The domain of the function comprises real numbers that correspond to angle measurements or arc lengths on the unit circle, while its range relates to the y-coordinates or sine values of the unit circle. Let us examine the unit circle seen in Figure 7.20. We shall imagine a line segment from the origin to a specific location on the circle as the pedal system of a bike, with the rider's foot positioned on the round itself. As the pedal undergoes rotation from an initial position of 0 radians along the horizontal axis to different angles, the associated angles (or arc lengths) and their respective heights from the horizontal axis, represented by the sine function sin, are utilized to generate data points on the graph of the sine function (Naidoo & Govender, 2014; Yu & Yu, 2020) (Figure 7.23).

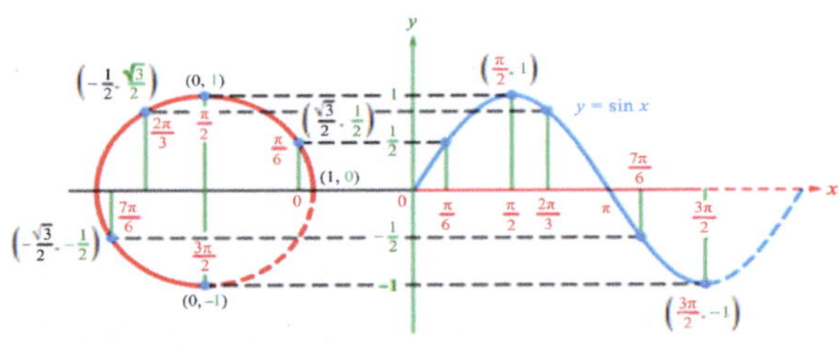

Figure 7.23. Sine graph showing bicycle example.

Source: Rob Hainer, Creative Commons License.

7.5.3. Graph of Cosine Function

The graph seen in Figure 7.24, which represents the function y = cos(x), has a similar form to that of the graph representing the function y = sin(x). The line graph of a cosine function may be seen as the graph of a sine function that has undergone a horizontal shift of π/2 units to the left (Çekmez, 2020; Irawan, Mukhlash, & Adzkiya, 2019).

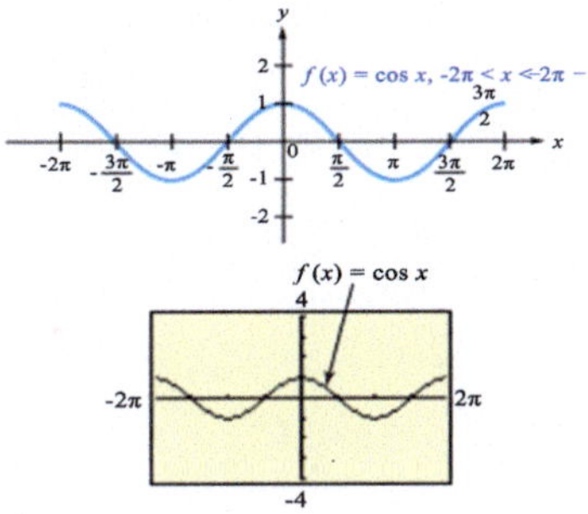

Figure 7.24. Cosine graph.

Source: Rob Hainer, Creative Commons License.

Here are the properties of the cosine function graph:

Domain: $(-\infty,\infty)$ $Range\left[-1,1\right]$

- The graph exhibits continuity over its full domain.
- The x-intercepts, or the intercepts, of the above function may be expressed in the following shape: (2n + 1) π/2, where n represents an integer.
- The period of the given function is 2π.
- The graph exhibits symmetry when compared to the y-axis, indicating that the function is classified as an even function.
- The equation cos(-x)= cos(x) is true for all values of x within the given domain.

The function f is said to be a function with even values if, for every value of x within the space of f, specific requirements are satisfied.

$$f\left(-x\right)=f\left(x\right)$$

An even function's graph exhibits symmetry concerning the y-axis. This implies that if a point (x, y) is a member of the function, then its reflection (-x, y) is likewise a member of the function. An illustration of the condition cos(-x)= cos(x) may be observed by considering the points (π/2, 0) and (-π/2, 0) on the graph of the function y= cos x, illustrating the property cos(-x) = cos x. (Mosese & Ogbonnaya, 2021).

The graphical representations of the function *f*= sin (x) in Figures 7.21 and 7.22, as well as the function *f*= cos(x) in Figure 7.24, are depicted inside a frame that spans approximately from -2π to 2π on the x-axis and from -4 to 4 on the y-axis. The x-axis scale (Xscl) is set to π/2, while the y-axis scale (Yscl) is set to 1. The present visual display represents the trigonometric display window (Maulyda et al., 2019; Rosjanuardi & Jupri, 2020).

The subsequent illustrations demonstrate graphs that undergo vertical or horizontal stretching or compression in comparison to the graphs of y= sin(x) or y= cos(x) (Setiawan, 2022).

- Example: Graphing y= asin(x)

  The graph of function y = 2sin(x) will now be examined and compared to the chart of the function y= sin(x).
- **Solution:** The graph of values demonstrates that for a specific amount of x, the related value of y is double the value it would be if y were equal to the sine of x. The sole modification observed in the chart is to the range, which is altered to encompass the values 2 and -2. Refer to Figure 7.25, which displays a graph of the function y= sin(x) for the sake of comparison.

x	0	$\frac{\pi}{2}$	π	$\frac{3\pi}{2}$	2π
$\sin x$	0	1	0	-1	0
$2 \sin x$	0	2	0	-2	0

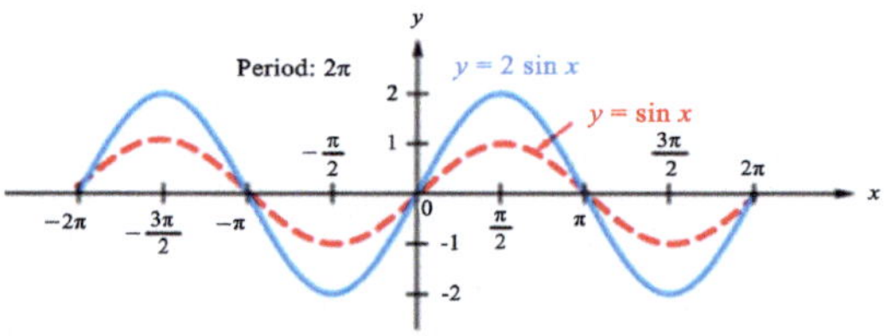

Figure 7.25. Graph of y=sin x.

Source: Lisa F., Creative Commons License.

The magnitude of a function with periodicity can be defined as one-half of the absolute difference between the highest and minimum values. The description pertains to the height of the graph, about a horizontal line that intersects the central point of the graph. Therefore, in the case of the fundamental sine value y = sin(x) (and likewise for the fundamental cosine function, which has the formula y= cos(x)), the amplitude is calculated in the following manner.

$$\frac{1}{2}[1-(-1)]=\frac{1}{2}(2)=1\,Amplitude\,of\ y=sinx$$

For y = 2 sin x, the amplitude is

$$\frac{1}{2}[2-(-2)]=\frac{1}{2}(4)=2\,Amplitude\,of\ y=2sinx$$

The graph $y = a\sin x$ may be conceptualized as a vertical expansion of the plot of $y = \sin x$ if the value of an is greater than 1, and as a vertical compression when the value of an is between 0 and 1.

Drawing a general conclusion from the aforementioned example yields the following:

The magnitude of the graph in equations y = asin(x) or y = acos(x) is influenced by value a, whereas the period is influenced by the coefficients of x in the argument. Let us examine the function $y = \sin 2x$. An Excel spreadsheet of values corresponding to the space between [0,2π] can be constructed.

x	0	$\frac{\pi}{4}$	$\frac{\pi}{2}$	$\frac{3\pi}{4}$	π	$\frac{5\pi}{4}$	$\frac{3\pi}{2}$	$\frac{7\pi}{4}$	2π
sin 2x	0	1	0	-1	0	1	0	-1	0

It is important to observe that a single full cycle takes place inside π units, rather than 2π units. Hence, the period in this context is represented by the mathematical constant π, which is equivalent to 2×π/2. Now let us consider the function $y = \sin 4x$.

x	0	$\frac{\pi}{8}$	$\frac{\pi}{4}$	$\frac{3\pi}{8}$	$\frac{\pi}{2}$	$\frac{5\pi}{8}$	$\frac{3\pi}{4}$	$\frac{7\pi}{8}$	π
sin 4x	0	1	0	-1	0	1	0	-1	0

These results indicate that a full cycle may be completed in π /2 or 2 × π/4 components.

$$\sin\left(4\cdot\frac{\pi}{2}\right)=sin2\pi=0$$

Typically, a graph of a parameter represented by the equations y= sin(bx) or y = cos(bx), where *b > 0*, will include an interval that is derived from 2π when *b* is not equal to 1. To comprehend the rationale behind this assertion, it is imperative to recall that the sine and cosine functions, sin(bx) and cos(bx), respectively, include the whole spectrum of potential values when bx traverses the interval between 0 to 2π. Hence, to determine the period of any one of these functions, it is necessary to solve the subsequent three-part inequality.

$$0 \le bx \le 2\pi$$

$$0 \le x \le \frac{2\pi}{b} \textit{divide each part by the positive number } b$$

Hence, the period of the function is calculated as 2π divided by *b*. By partitioning the range [0, 2π/b] into four equidistant segments, we may determine the values at which the trigonometric functions sin(bx) or cos(bx) attain the values of -1, 0, or 1. These values will yield minimum indications, x-intercepts, and the highest locations on the graph. After establishing these locations, the graph may be constructed by connecting them using a continuous sinusoidal curve. If a function possesses a value of *b* that is less than zero, it is possible to utilize the identities discussed in the subsequent chapter to reframe the function in such a way that *b* becomes greater than zero.

7.5.4. Translation of the Graphs of Sine and Cosine Functions

7.5.4.1. Horizontal Translation

The graph of the function:

$$y = f(x - d)$$

is horizontally shifted in comparison to the one of y= *f*(x). The displacement is *d* units to the opposite direction if *d* is more than 0. While the displacement is *d* units to the left if *d* is less than 0 (Hidalgo et al., 2020; Boulton et al., 2021).

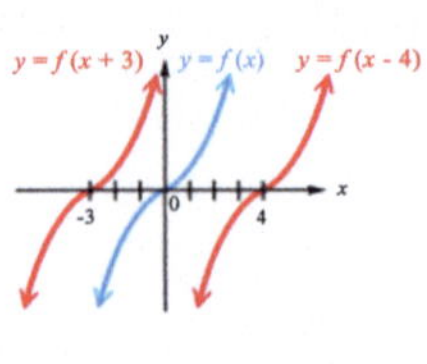

Figure 7.26. Horizontal translations of y=f (x).

Source: Lisa F., Creative Commons License.

In the context of circular functions, the process of horizontally translating a function is referred to as a phase shift. The parameter in the equation for y = *f*(x – d) is represented by the phrase *x – d* (Figure 7.26).

7.5.4.2. Vertical Translations

The graph of a function:

$$y = c + f(x)$$

is vertically translated compared to y = f (x). The transformation is *c* units up if *c* is more than 0, and is the absolute value of *c* units down if *c* is less than 0 (Yuan et al., 2021; Sun, 2022) (Figure 7.27).

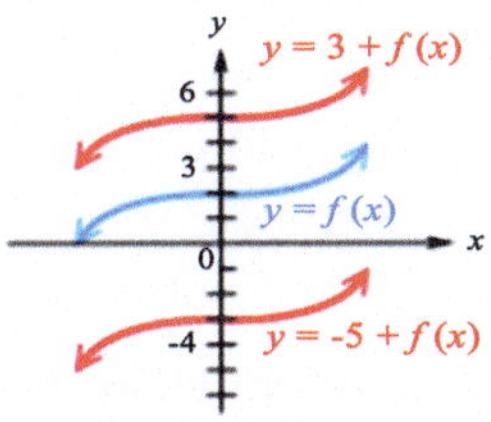

Figure 7.27. Vertical translations of y=f(x).

Source: Lisa F., Creative Commons License.

7.5.4.3. Combination of Translations

A function of the form:

$$y = c + a\sin\left[b\left(x-d\right)\right] or\ y = c + a\cos\left[b\left(x-d\right)\right], where\ b > 0$$

which encompasses stretching, shrinking, and translating, may be executed according to the criteria explained in the following sections (Bongardt, 2020; Hidalgo et al., 2020).

7.5.5. Further Guidelines for Sketching Graphs of Sine and Cosine Functions

Method I: Please adhere to the following instructions:

- By solving the three-part inequality 0 < b(x – d) ≤ 2π, we may find an interval with a length equal to one period 2π/2b.
- The interval should be divided into four equal halves. Compute the value of the formula for each of the x-values obtained.

- The points under consideration will consist of the highest locations, lowest points, and points that intersect with the line y = c, which can be referred to as the "middle" points of the wave.
- The points obtained in Step 3 should be plotted and connected by a sinusoidal curve characterized by an amplitude denoted as 'a.'
- Extend the graph to include new periods, if required.

Method II: Please adhere to the following instructions:

The equations to be graphed are y = asin(bx) and y = acos(bx).

The magnitude of the function's amplitude is represented as |a|, and the duration of one complete cycle is denoted as 2π/b.

Utilize translation techniques to visually represent the intended mathematical function on a graph.

In the context of vertical translation, it can be observed that a positive value of *c* results in an upward shift of *c* units, whereas a negative value of *c* leads to a downward shift of the absolute value of *c* units (Joshi, 2022).

The horizontal translation, also known as the phase shift, is *d* units to the right when *d* is more than 0, and it is |d| units to the left when *d* is less than 0 (Mahmood et al., 2022).

7.5.6. Graph of Tangent Function

Refer to the table that lists the chosen points provided with the graphical representation of the tangent function shown in Figure 7.28. These points encompass the range of exceptional values that lie between -π/2 and π/2. The tangent function has undefined behavior for odd multiples of π/2, resulting in the presence of vertical asymptotes at these particular values. The vertical asymptote is defined as an imaginary line that the graph of a function approaches but fails to intersect. As the independent variable approaches the line, the dependent variable exhibits unbounded growth or decline (Kocayusufoğlu, 2000; Orhani, 2022). Moreover, given that:

$$\tan(-x) = -tanx$$

The period of the tangent function is π.

x	$y = \tan x$
$-\frac{\pi}{3}$	$-\sqrt{3} \approx -1.7$
$-\frac{\pi}{4}$	-1
$-\frac{\pi}{6}$	$-\frac{\sqrt{3}}{3} \approx -0.6$
0	0
$\frac{\pi}{6}$	$\frac{\sqrt{3}}{3} \approx 0.6$
$\frac{\pi}{4}$	1
$\frac{\pi}{3}$	$\sqrt{3} \approx 1.7$

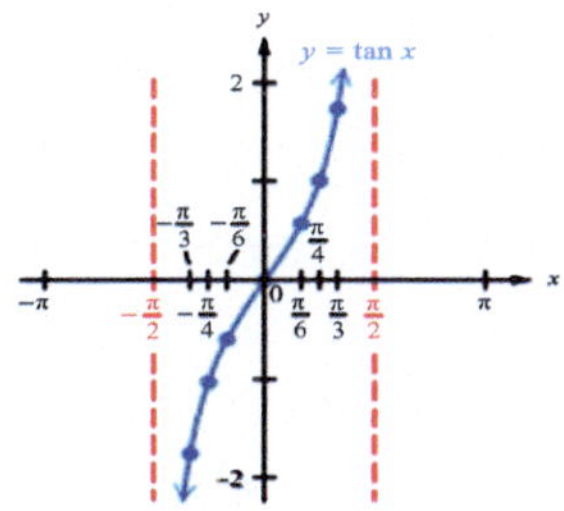

Figure 7.28. Graph with a table of tangent functions.

Source: Alexander, Creative Commons License.

The equation tan(x)= sin(x)/cos(x) implies that the tangent values are equal to zero when the sine values are zero and are undefined when the cosine values are zero. As the x-values climb from -π/2 to π/2, the tangent values span from negative infinity to positive infinity and exhibit a continuous increase over the whole interval. The aforementioned numbers exhibit repetition when x progressively grows from π/2 to 3π/2, from 3π/2 to 5π/2, and so on. Figure 7.29 illustrates the graphical representation of the function y = tan(x) across the interval from – 3π/2 to 3π/2 (Karpfinger, 2022). The graph (Figure 7.29) exhibits a consistent pattern of continuation.

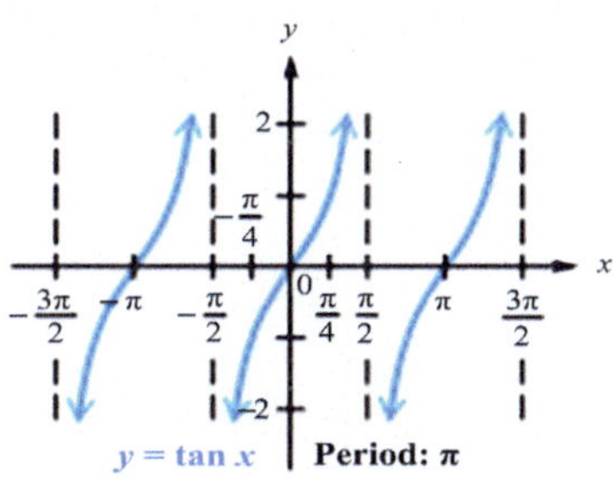

Figure 7.29. Graph of y=tan x.

Source: Alexander, Creative Commons License.

7.5.7. Graph of Cotangent Function

The graph in Figure 7.30 depicts the results of a comparative study conducted on specific data points ranging from 0 to π for the cotangent function (Kocayusufoğlu, 2000; Sajid, 2020). The vertical asymptotes in this context occur at x-values, which are multiples of an integer of π. Because

$$\cot(-x) = -cotx$$

The graph exhibits symmetry concerning its origin. This phenomenon becomes evident when a greater portion of the graph is displayed.

x	$y = \cot x$
$\frac{\pi}{6}$	$\sqrt{3} \approx 1.7$
$\frac{\pi}{4}$	1
$\frac{\pi}{3}$	$\frac{\sqrt{3}}{3} \approx 0.6$
$\frac{\pi}{2}$	0
$\frac{2\pi}{3}$	$-\frac{\sqrt{3}}{3} \approx -0.6$
$\frac{3\pi}{4}$	-1
$\frac{5\pi}{6}$	$-\sqrt{3} \approx -1.7$

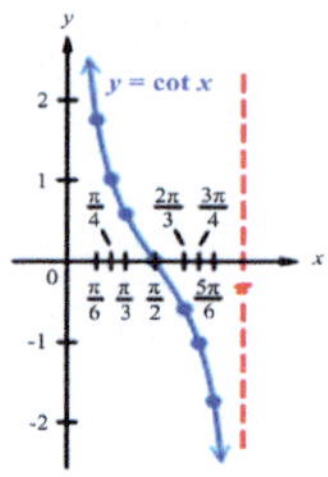

Figure 7.30. Graph for cotangent function.

Source: Alexander, Creative Commons License.

The cotangent function exhibits a periodicity of π. The cotangent values are equal to zero when the cosine values are equal to zero, and they are undefined when the sine values are equal to zero. As the x-values climb from 0 to π, the cotangent values exhibit a range from positive infinity to negative infinity, and they consistently decrease within this interval. The aforementioned numbers exhibit repetition when the variable *x* progressively grows from π to 2π, from 2π to 3π, and so on. Figure 7.31 illustrates the graph of the function y = cot(x) over the interval from negative π to positive π (SUSAM, 2020). The graph exhibits a consistent continuation of the observed trend.

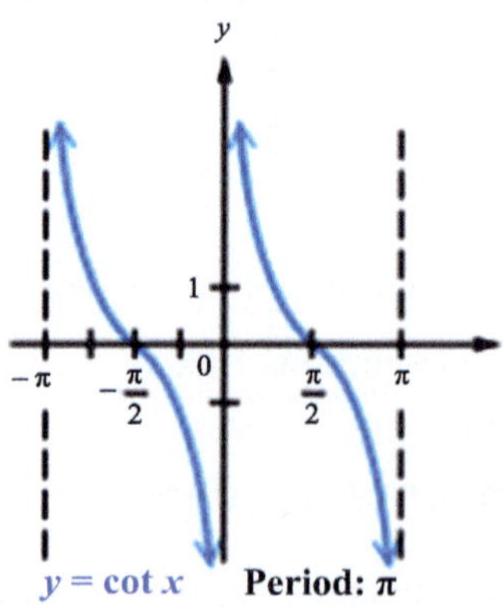

Figure 7.31. Graph of y=cotx.

Source: Harry Spurling, Creative Commons License.

The graph of the tangent function may be conveniently plotted on a graphing calculator by utilizing the dedicated tangent key. To plot the cotangent function on a graph, it is necessary to utilize one of the established identities, as most graphing calculators do not have direct access to the cotangent function.

$$\cot x = \frac{1}{tanx} or\ cotx = \frac{cosx}{sinx}$$

7.5.8. Guidelines for Sketching Graphs of Tangent and Cotangent Function

To graph $y = a\tan(bx) or\ y = a\cot(bx)$, with b > 0, follow these steps:

- Determine the period: $\frac{\pi}{b}$. To locate two adjacent vertical asymptotes, solve the following equations for x:

  $for\ y = a\tan(bx); bx = -\pi/2\, and\ bx = \pi/2$

  $for\ y = \text{acot}(bx); bx = 0\, and\ bx = \pi$
- Sketch the two asymptotes identified in Step 1.
- To proceed, it is necessary to divide the gap that is produced by the asymptotes into four equal portions.
- To assess the function, it is necessary to compute its values at the first-quarter point, midway, and third-quarter point, utilizing the x-values obtained in Step 3.
- The task at hand involves connecting the given points using a continuous curve that exhibits a gradual approach toward the vertical asymptotes. Identify any other asymptotes and periods in the graph, if applicable.

7.6. GRAPHS OF THE SECANT & COSECANT FUNCTIONS

7.6.1. Graph of the Secant Function

Please refer to the table containing the chosen points provided with the graphical representation of the secant function, shown in Figure 7.32. These points encompass a range of exceptional values spanning from negative pi to positive pi (Fulling, 2005; Hertel & Cullen, 2011). Moreover, given that:

$$\sec(-x) = \sec x$$

The graph that represents the secant function exhibits symmetry compared to the y-axis.

x	$y = \sec x$
0	1
$\pm\frac{\pi}{6}$	$\frac{2\sqrt{3}}{3} \approx 1.2$
$\pm\frac{\pi}{4}$	$\sqrt{2} \approx 1.4$
$\pm\frac{\pi}{3}$	2
$\pm\frac{2\pi}{3}$	-2
$\pm\frac{3\pi}{4}$	$-\sqrt{2} \approx -1.4$
$\pm\frac{5\pi}{6}$	$-\frac{2\sqrt{3}}{3} \approx -1.2$
$\pm\pi$	-1

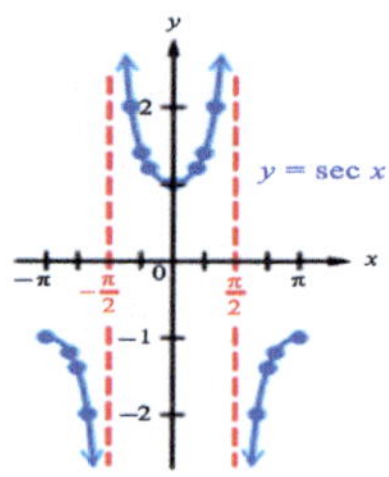

Figure 7.32. Graph of secant function.

Source: Paul Mathew, Creative Commons License.

The period of the secant function is 2π, which is equivalent to the period of the cosine function, y = cos(x), due to the reciprocal relationship between secant values and corresponding cosine values. When the cosine of *x* equals 1, it follows that the secant of *x* is likewise equal to 1. Similarly, in the case when the value of cosine *x* is equal to -1, the value of secant *x* is also -1. For every value of *x*, the cosine function is bounded between -1 and 1, indicating that the absolute value of the secant function is always greater than or equal to 1 for any *x* within its domain (Wilson, 2008). Figure 7.33 depicts the interrelationship between the graphs of the trigonometric functions y = cos(x) and y = sec(x).

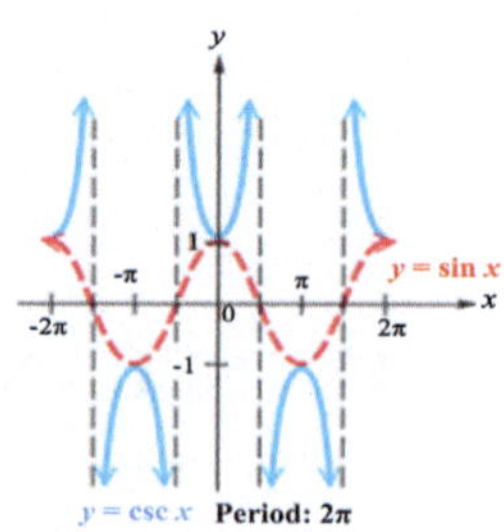

Figure 7.33. Graph of y=cos x.

Source: Paul Mathew, Creative Commons License.

7.6.2. Graph of Cosecant Function

The graph in Figure 7.34 depicts the results of a comparative study conducted on certain points within the range of -π to π for the graphical representation of the cosecant function (Thompson & Dray, 2000). The x-values at which the vertical asymptotes occur are multiples of an integer of π.

The graph exhibits symmetry concerning the point of origin.

$\csc(-x) = -cscx$

x	$y = \csc x$	x	$y = \csc x$
$\frac{\pi}{6}$	2	$-\frac{\pi}{6}$	-2
$\frac{\pi}{4}$	$\sqrt{2} \approx 1.4$	$-\frac{\pi}{4}$	$-\sqrt{2} \approx -1.4$
$\frac{\pi}{3}$	$\frac{2\sqrt{3}}{3} \approx 1.2$	$-\frac{\pi}{3}$	$-\frac{2\sqrt{3}}{3} \approx -1.2$
$\frac{\pi}{2}$	1	$-\frac{\pi}{2}$	-1
$\frac{2\pi}{3}$	$\frac{2\sqrt{3}}{3} \approx 1.2$	$-\frac{2\pi}{3}$	$-\frac{2\sqrt{3}}{3} \approx -1.2$
$\frac{3\pi}{4}$	$\sqrt{2} \approx 1.4$	$-\frac{3\pi}{4}$	$-\sqrt{2} \approx -1.4$
$\frac{5\pi}{6}$	2	$-\frac{5\pi}{6}$	-2

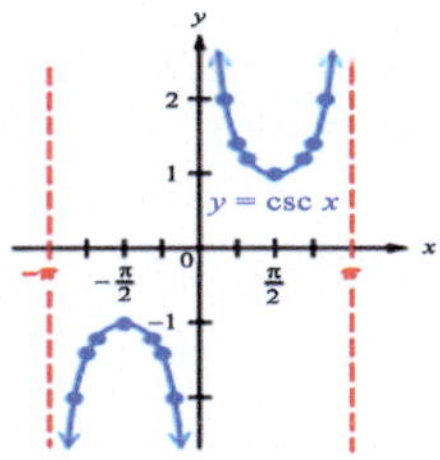

Figure 7.34. Graph of cosecant function.

Source: Paul Mathew, Creative Commons License.

The period of the cosecant function is 2π, which is equivalent to the period of the sine function due to the reciprocal relationship between cosecant values and their corresponding sine values. When the sine of *x* is equal to 1, the cosecant of *x* is likewise equal to 1. Similarly, in the case when sin(x) equals -1, the reciprocal of sin(x), known as csc(x), also equals -1. For every value of *x*, the sine function is bounded between -1 and 1, which implies that the reciprocal of the cosecant function, denoted as csc(x), is greater than or equal to 1 for any *x* within its domain. The relationship between the two charts of y = sin(x) and y = csc(x) is seen in Figure 7.35.

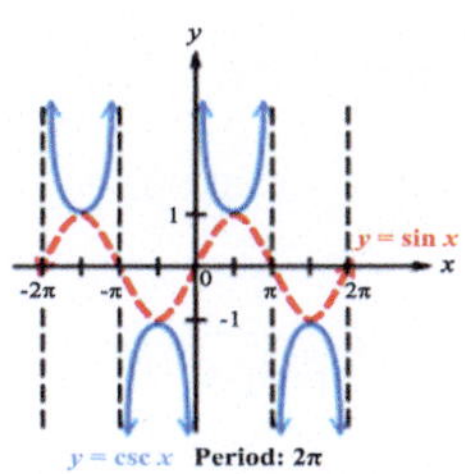

Figure 7.35. Graph of y=csc x.

Source: J. Bell, Creative Commons License.

In general, it is common for calculators to lack dedicated keys to perform cosecant and secant trigonometric functions. To plot the graph of the function y = csc(x) using a graphing calculator, one needs to employ the following steps.

$$cscx = \frac{1}{sinx} Reciprocal\ Identity$$

In Figure 7.36, the chart of Y_1 = sin(X) is shown by a thin line, whereas the graph of Y_2 = csc(X) is represented by a thicker line (Fogiel et al., 1984). The calculator display does not provide information on the vertical asymptotes. However, it is important to note that the vertical asymptotes of the guide functional Y_1 = sin(X) are located at each x-intercept (Azim, Akhi, & Kamrujjaman, 2023; Ninness et al., 2009). The graph depicted in Figure 7.37 illustrates the secant function, which has been plotted in a manner that is analogous to the given identity.

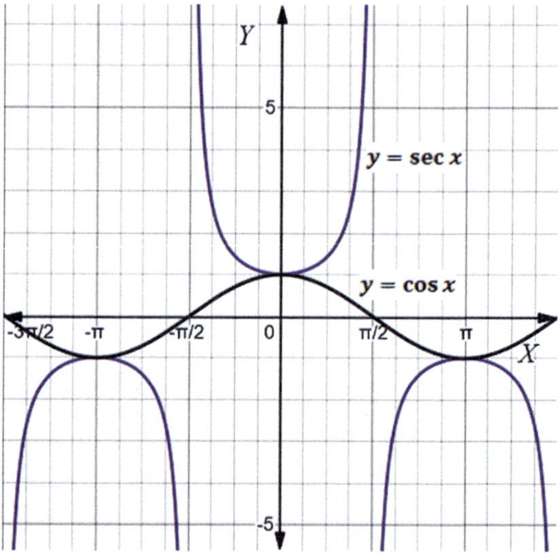

Figure 7.36. Graph Y_1=Cos X and Y_2=sec X.

Source: J. Bell, Creative Commons License.

$$Reciprocal\ Identity : \sec X = \frac{1}{\cos X}$$

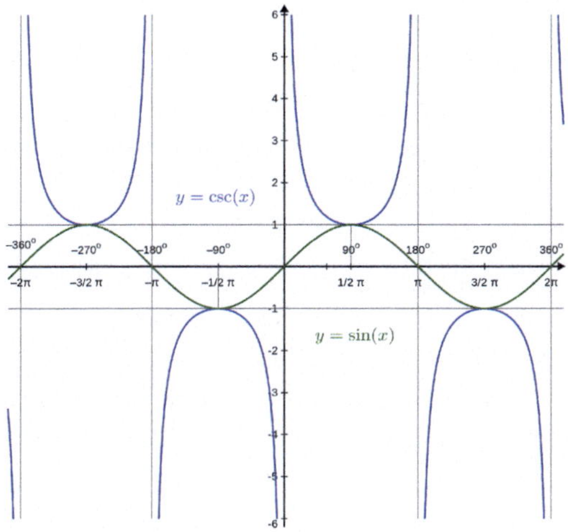

Figure 7.37. Graph of Y_1=sin X & Y_2=csc X.
Source: J. Bell, Creative Commons License.

7.6.3. Guidelines for Sketching Graphs of Cosecant and Secant Functions

To plot the functions y = acsc(bx) or y = asec(bx), where b > 0, the following procedure should be followed.

1. To provide a visual reference, it is recommended to graph the reciprocal function with a dashed curve.

To Graph	Use as a Guide
$y = a\csc(bx)$	$y = a$
$y = a\sec(bx)$	$y = a$

2. Sketch the vertical asymptotes. The equations will be in the form of x = k, where k represents an x-intercept on the graph of the reference function.
3. Sketch the chart of the desired functions by depicting the characteristic U-shaped branches connecting the neighboring asymptotes.

Similar to the graphical representations of the sine and cosine functions, the graphs representing the secant and cosecant functions are subject to vertical and horizontal translations. The period of both fundamental functions remains 2π (Fotiyev & Fotiyeva, 2012; Van Brummelen, 2012).

7.7. SUMMARY

The chapter about Mathematical Functions and Graphs explores the fundamental principles of trigonometry, with a specific emphasis on the trigonometric equations of degrees and their graphic depictions. Trigonometry is a mathematical discipline concerned with the interconnections among the sides and angles of a triangle. The utilization of this phenomenon spans several disciplines like mathematics, engineering, and astronomy.

The chapter starts by providing an introduction to the trigonometric functions associated with an angle, encompassing sine, cosine, tangent, cotangent, secant, as well as cosecant. The aforementioned functions are established based on the proportions of the sides of a triangle with a right angle, serving as a means to establish a connection between the angles as well as the sides of such a triangle. The central emphasis lies in the examination of the interconnections among these functions and their application in resolving diverse problems about angles and distances.

The chapter extensively discusses the differentiation among trigonometric functions of acute angles compared to those of non-acute angles. Acute angles are defined as angles that measure below ninety degrees, whereas non-acute angles encompass both obtuse angles (which measure greater than 90 degrees) and straight angles $(180°)$. The examination of the characteristics and actions of trigonometric functions is conducted in many situations, with particular emphasis on the alterations in their numerical outputs when angles extend beyond the realm of acute angles.

Extensive discourse is dedicated to the analysis of graphical depictions of trigonometric functions. The sinusoidal curves, often known as the graphs of cosine and sine functions, demonstrate periodic characteristics by oscillating between the values of -1 and 1 while the angle undergoes variation. The aforementioned functions exhibit symmetry about the origin and display an endless repeating pattern.

The chapter also discusses the graphical representations of tangential and cotangent functions. The periodic character of these functions results in the presence of asymptotes at regular intervals. The graphs of tangent and cotangent functions exhibit periodic behavior characterized by vertical dilations and compressions.

In addition, an examination is conducted on the graphical representations of secant along with cosecant functions. The graphs under consideration exhibit resemblances to their cosine and sine counterparts, although they manifest vertical asymptotes in instances when the value of cosine or sine functions approaches zero. The graphs of the secant and cosecant functions also demonstrate periodic patterns and fluctuations in amplitude.

REFERENCES

1. Adamek, T., Penkalski, K., & Valentine, G., (2005). The history of trigonometry. *History of Mathematics, 11.*
2. Agarwal, R. P., (2020). Pythagorean theorem before and after Pythagoras. *Advanced Studies in Contemporary Mathematics, 30*(3), 357–389.
3. Ali, M. N., Schoop, L. M., Garg, C., Lippmann, J. M., Lara, E., Lotsch, B., & Parkin, S. S., (2016). Butterfly magnetoresistance, quasi-2D Dirac Fermi surface and topological phase transition in ZrSiS. *Science Advances, 2*(12), e1601742.
4. Arts, T., Prinzen, F. W., Delhaas, T., Milles, J. R., Rossi, A. C., & Clarysse, P., (2010). Mapping displacement and deformation of the heart with local sine-wave modeling. *IEEE Transactions on Medical Imaging, 29*(5), 1114–1123.
5. Azim, M., Akhi, A. A., & Kamrujjaman, M., (2023). *A Series of New Formulas to Approximate the Sine and Cosine Functions, 1, 1-19.*
6. Bond, J. D., (1921). The development of trigonometric methods down to the close of the XVth century. *ISIS, 4*(2), 295–323.
7. Bongardt, B., (2020). The adjoint trigonometric representation of displacements and a closed-form solution to the IKP of general 3C chains. *ZAMM-Journal of Applied Mathematics and Mechanics/ Zeitschrift für Angewandte Mathematik und Mechanik, 100*(6), e201900214.
8. Bornstein, N., (2020). Teaching transformations of trigonometric functions with technology. *Journal of Interactive Media in Education, 1*(15), 1–9.
9. Boulton, L., Olver, P. J., Pelloni, B., & Smith, D. A., (2021). New revival phenomena for linear integro–differential equations. *Studies in Applied Mathematics, 147*(4), 1209–1239.
10. Bressoud, D. M., (2010). Historical reflections on teaching trigonometry. *The Mathematics Teacher, 104*(2), 106–112.
11. Calzada, M. E., & Scariano, S. M., (2006). Activities for students: A natural bridge from algebra and geometry to trigonometry. *The Mathematics Teacher, 99*(6), 450–458.
12. Cavey, L. O., & Berenson, S. B., (2005). Learning to teach high school mathematics: Patterns of growth in understanding right triangle trigonometry during lesson plan study. *The Journal of Mathematical*

Behavior, 24(2), 171–190.

13. Çekmez, E., (2020). What generalizations do students achieve with respect to trigonometric functions in the transition from angles in degrees to real numbers? *The Journal of Mathematical Behavior, 58*, 100778.
14. Cetin, O. F., (2015). Students perceptions and development of conceptual understanding regarding trigonometry and trigonometric function. *Educational Research and Reviews, 10*(3), 338–350.
15. Challenger, M., (2009). *From Triangles to a Concept: A Phenomenographic Study of A-Level Students' Development of the Concept of Trigonometry.* University of Warwick,
16. Chin, K. E., & Tall, D., (2012). *Making Sense of Mathematics Through Perception, Operation and Reason: The Case of Trigonometric Functions.* Paper presented at the Proceedings of the 36th Conference of the International Group for the Psychology of Mathematics Education.
17. Curri, E., (2012). *Using Computer Technology in Teaching and Learning Mathematics in an Albanian Upper Secondary School: The Implementation of SimReal in Trigonometry Lessons.* Universitetet i Agder; University of Agder,
18. DeJarnette, A. F., (2018). Students' conceptions of sine and cosine functions when representing periodic motion in a visual programming environment. *Journal for Research in Mathematics Education, 49*(4), 390–423.
19. Demir, Ö., & Heck, A., (2013). *A new Learning Trajectory for Trigonometric Functions.* Paper presented at the Proceedings of the 11th international conference on technology in mathematics teaching.
20. Dündar, S., (2015). Mathematics teacher-candidates' performance in solving problems with different representation styles: The trigonometry example. *Eurasia Journal of Mathematics, Science and Technology Education, 11*(6), 1379–1397.
21. Fogiel, M., Research, & Association, E., (1984). *Handbook of Mathematical, Scientific, and Engineering Formulas, Tables, Functions, Graphs, Transforms*. Research & Education Assoc.
22. Fotiyev, D., & Fotiyeva, I., (2012). *The Complete Idiot's Guide to Trigonometry: Master Trig with Crystal-Clear Explanations of all the Basics*. Dorling Kindersley Ltd.
23. Fulling, S. A., (2005). How to avoid the inverse secant (and even the

secant itself). *The College Mathematics Journal, 36*(5), 381–387.

24. Garibaldi, S., (2008). Somewhat more than governors need to know about trigonometry. *Mathematics Magazine, 81*(3), 191–200.
25. Guirguis, M., (2019). *Introduction to Discrete Mathematics and Fourier Analysis. A Practical Guide for Third Year Undergraduate Students.* A practical guide for third year undergraduate students.
26. Han, X., (2015). Normalized B-basis of the space of trigonometric polynomials and curve design. *Applied Mathematics and Computation, 251*, 336–348.
27. Hertel, J., & Cullen, C., (2011). *Teaching Trigonometry: A Directed Length Approach.* Paper presented at the Proceedings of the 33rd annual MEETING of the North American chapter of the international group for the psychology of mathematics education.
28. Hidalgo, J. I., Botella, M., Velasco, J. M., Garnica, O., Cervigón, C., Martínez, R., . Lanchares, J., (2020). Glucose forecasting combining Markov chain based enrichment of data, random grammatical evolution and bagging. *Applied Soft Computing, 88*, 105923.
29. Irawan, M. I., Mukhlash, I., & Adzkiya, D., (2019). *Development of Trigonometric Visualization Concepts to Increase the Study Motivations of SMK Students.* Paper presented at the Journal of Physics: Conference Series.
30. Joshi, R. R., (2022). *Students' Achievement in Trigonometry Teaching Through Geogebra.* Department of Mathematics Education,
31. Kalanov, T. Z., (2014). On the system analysis of the foundations of trigonometry. *Bulletin of Pure & Applied Sciences-Mathematics and Statistics, 33*(1), 1–27.
32. Kamber, D., & Takaci, D., (2018). On problematic aspects in learning trigonometry. *International Journal of Mathematical Education in Science and Technology, 49*(2), 161–175.
33. Karpfinger, C., (2022). Trigonometric functions. In: *Calculus and Linear Algebra in Recipes: Terms, Phrases and Numerous Examples in Short Learning Units* (pp. 49–58): Springer.
34. Kendal, M., & Stacey, K., (1996). *Trigonometry: Comparing Ratio and Unit Circle Methods.* Paper presented at the Technology in mathematics education. Proceedings of the 19th annual conference of the Mathematics Education Research Group of Australasia.
35. King, C. D., Evelyn, T., Ye, F., & Carvajal, B. E., (2020). *Trigonometry:*

A Brief Conversation, 1-1-10.

36. Kocayusufoğlu, İ., (2000). Trigonometry on iso-taxicab geometry. *Mathematical and Computational Applications, 5*(3), 201–212.
37. Koks, D., (2022). Wanted: A name for a commonly used trigonometry function. *The Journal of Engineering, 2022*(12), 1143–1148.
38. Liu, T., Mou, J., Xiong, L., Han, X., Yan, H., & Cao, Y., (2021). Hyperchaotic maps of a discrete memristor coupled to trigonometric function. *Physica Scripta, 96*(12), 125242.
39. Liu, Y., Liu, B., Liu, Y., & Drew, M. G., (2011). Correlations between two sets of angular relation equations. *Journal of Mathematical Chemistry, 49*, 2089–2108.
40. Mahmood, Z., Khogeer, H. A., Hussam, H. E., & Hossain, M. M., (2022). A logistic trigonometric generalized class of distribution characteristics, applications, and simulations. *Journal of Mathematics, 2022.*
41. Maknun, C., Rosjanuardi, R., & Jupri, A., (2019). *From Ratios of Right Triangle to Unit Circle: An Introduction to Trigonometric Functions.* Paper presented at the Journal of Physics: Conference Series.
42. Martínez-Planell, R., & Delgado, A. C., (2016). The unit circle approach to the construction of the sine and cosine functions and their inverses: An application of APOS theory. *The Journal of Mathematical Behavior, 43*, 111–133.
43. Masseran, N., Razali, A. M., Ibrahim, K., Zaharim, A., & Sopian, K., (2015). *Fitting a Circular Distribution Based on Nonnegative Trigonometric Sums for Wind Direction in Malaysia.* Paper presented at the AIP Conference Proceedings.
44. Maulyda, M. A., Hidayanto, E., & Rahardjo, S., (2019). Representation of trigonometry graph funcsion colage students using GeoGebra. *International Journal of Trends in Mathematics Education Research, 2*(4), 193–196.
45. May, V., & Courtney, S., (2016). Developing meaning in trigonometry. *Illinois Mathematics Teacher, 63*(1), 25–33.
46. Mokhtar, A., Ayub, M., Ismail, N., & Daud, N., (2018). *Implementation of Trigonometric Function Using CORDIC Algorithms.* Paper presented at the AIP Conference Proceedings.
47. Moore, F. R., (1978). An introduction to the mathematics of digital signal processing: Part I: Algebra, trigonometry, and the most beautiful

formula in mathematics. *Computer Music Journal,* 38–47.

48. Moore, K. C., (2010). *The Role of Quantitative Reasoning in Precalculus Students Learning Central Concepts of Trigonometry.* Arizona State University.
49. Moore, K. C., (2012). Coherence, quantitative reasoning, and the trigonometry of students. *Quantitative Reasoning and Mathematical Modeling: A Driver for STEM Integrated Education and Teaching in Context, 2*, 75–92.
50. Moritz, R. É., (1913). *Plane and Sperical Trigonometry (with Five-Place Tables): A Text-book for Technical Schools and Colleges.* J. Wiley.
51. Mosese, N., & Ogbonnaya, U. I., (2021). GeoGebra and students' learning achievement in trigonometric functions graphs representations and interpretations. *Cypriot Journal of Educational Sciences, 16*(2), 827–846.
52. Naidoo, J., & Govender, R., (2014). Exploring the use of a dynamic online software program in the teaching and learning of trigonometric graphs. *Pythagoras, 35*(2), 1–13.
53. Ninness, C., Dixon, M., Barnes-Holmes, D., Rehfeldt, R. A., Rumph, R., McCuller, G., . McGinty, J., (2009). Constructing and deriving reciprocal trigonometric relations: A functional analytic approach. *Journal of Applied Behavior Analysis, 42*(2), 191–208.
54. Orhani, S., (2022). Performing homework using of ICT for graphical presentation of trigonometric functions: A qualitative analysis. *International Research Journal of Science, Technology, Education, & Management (IRJSTEM), 2*(2).
55. Price, C., & Van, J. P., (2017). Using open-response tasks to reveal the conceptual understanding of learners-learners teaching the teacher what they know about trigonometry. *African Journal of Research in Mathematics, Science and Technology Education, 21*(2), 159–175.
56. Raychev, N., (2015). Bilaterally symmetrical transformation between independent operators and rotations. *Journal of Quantum Information Science, 5*(3), 79.
57. Rosjanuardi, R., & Jupri, A., (2020). Didactical design on drawing and analyzing trigonometric functions graph through a unit circle approach. *International Electronic Journal of Mathematics Education, 15*(3).
58. Sajid, M., (2020). Singular values and real fixed points of one-parameter

families associated with fundamental trigonometric functions\\[4pt] $\sin z$, $\cos z$ and $\tan z$. *International Journal of Applied Mathematics, 33*(4), 635.

59. Sanna, A., Reid, M., Dame, T., Menten, K., Brunthaler, A., Moscadelli, L., & Xu, Y., (2011). Trigonometric parallaxes of massive star-forming regions. IX. The outer arm in the first quadrant. *The Astrophysical Journal, 745*(1), 82.
60. Setiawan, Y. E., (2022). Prospective teachers representations in problem solving of special angle trigonometry functions based on the level of ability. *Infinity Journal, 11*(1), 55–76.
61. Stitz, C., & Zeager, J., (2011). *College Trigonometry*. Stitz Zeager Open-Source Mathematics.
62. Sun, Z., (2022). A conservative scheme for two-dimensional Schrödinger equation based on multiquadric trigonometric quasi-interpolation approach. *Applied Mathematics and Computation, 423*, 126996.
63. Susam, S. O., (2020). A new family of Archimedean copula via trigonometric generator function. *Gazi University Journal of Science,* 1.
64. Thompson, K., & Dray, T., (2000). Taxicab angles and trigonometry. *Pi Mu Epsilon Journal,* 87–96.
65. Ungar, A. A., (2001). Hyperbolic trigonometry and its application in the Poincaré ball model of hyperbolic geometry. *Computers & Mathematics with Applications, 41*(1, 2), 135–147.
66. Van, B. G., (2012). *Heavenly Mathematics: The Forgotten Art of Spherical Trigonometry*. Princeton University Press.
67. Wallis, J., (1974). *A Treatise of Algebra, Both Historical and Practical: Shewing the Original, Progress, and Advancement Thereof, from Time to Time: And by What Steps it Hath Attained to the Heighth at Which it Now is. With Some Additional Treatises, I. Of the Cono-cuneus. II. Of Angular Sections; and other Things Relating Thereunto, and to Trigonometry. III. Of the Angle of Contact. IV. Of Combinations, Alternations, and Aliquot Parts, 1, 1-19.*
68. Walsh, R., Fitzmaurice, O., & O'Donoghue, J., (2017). What subject matter knowledge do second-level teachers need to know to teach trigonometry? An exploration and case study. *Irish Educational Studies, 36*(3), 273–306.

69. Weber, K., (2008). Connecting research to teaching: Teaching trigonometric functions: Lessons learned from research. *The Mathematics Teacher, 102*(2), 144–150.
70. Wilson, S. J., (2008). Dynamic web tools for trigonometry. *International Journal of Continuing Engineering Education and Life Long Learning, 18*(5, 6), 698–707.
71. Yahya, N. A., Varatharajoo, R., & Harithuddin, A., (2020). Satellite formation flying relative geodesic and latitudinal error measures. *Journal of Aeronautics, Astronautics and Aviation, 52*(1), 83–93.
72. Yang, D. C., & Sianturi, I. A., (2017). An analysis of Singaporean versus Indonesian textbooks based on trigonometry content. *Eurasia Journal of Mathematics, Science, and Technology Education, 13*(7), 3829–3848.
73. Yao, Z., Cui, X., Lu, M., Feng, Z., & Yang, J., (2010). Pseudo-correlation-function-based unambiguous tracking technique for sine-BOC signals. *IEEE Transactions on Aerospace and Electronic Systems, 46*(4), 1782–1796.
74. Yeshurun, S., (1982). The angle: A logical gap in teaching geometry and trigonometry and its remedy. *International Journal of Mathematical Education in Science and Technology, 13*(2), 133–138.
75. Yu, W., & Yu, T., (2020). Analysis of chaotic characteristics of trigonometric function system. *Modern Physics Letters B, 34*(21), 2050210.
76. Yuan, L., BAO, S., Yong-chao, L., & Xiu-hong, C., (2021). Research on Mongolian-Chinese translation model based on transformer with soft context data augmentation technique. *IEICE Transactions on Fundamentals of Electronics, Communications and Computer Sciences*.
77. Zainal, B., & Ismail, M. R., (2010). Trigonometric solutions using sine quadrant. *Procedia-Social and Behavioral Sciences, 8*, 721–728.

CHAPTER 8

Trigonometry and Technology

Contents

8.1. INTRODUCTION

Trigonometry fundamentally exemplifies human intellectual prowess, originating from the inherent inquisitiveness that prompted early thinkers to contemplate the interconnections between heavenly bodies and earthly topographies. Trigonometry was initially endowed with its purpose through the endeavors of astronomers who recorded the celestial movements of stars, as well as the efforts of builders who constructed structures with meticulous accuracy (Curri, 2012). Over the course of thousands of years, the concepts of this subject have undergone development, resulting in the emergence of a mathematical language consisting of proportions, processes, and constants that convey a universally applicable truth transcending temporal and spatial boundaries (Qiao et al., 2022).

However, in the contemporary period, trigonometry's significance is further emphasized, as it is now enhanced by the integration of advanced technological tools. Computers have provided us with the computing power necessary to change these long-standing ratios with an unparalleled level of precision (Zengin et al., 2012). Algorithms govern these mathematical operations, giving rise to both digital artwork and impressive feats of engineering. The abstract notions of trigonometry are now being represented through pixel-perfect animations, realistic simulations, and fluid movements (Jelatu et al., 2019).

Trigonometry has evolved beyond its geometric origins and has found its way into different areas of science and technology. It serves as a useful link between abstract mathematical concepts and everyday problem-solving in disciplines such as science, engineering, navigation, computer graphics, electronics, music, and emerging technologies.

Trigonometry has made a significant impact on several aspects of human civilization, as seen by the towering buildings that reach for the sky, the gravity-defying arching bridges, and circuits that power and enhance our daily lives (Johnson & Walker, 2011). Surveying professionals, designers, and engineers utilize the power of these principles to transform conceptual designs into physical realities, drawing upon the knowledge of distances and angles that have been passed down through generations (Serpe & Frassia, 2021).

Trigonometry assumes a critical role in navigation inside the digital domains of satellite and GPS, allowing precise localization on a worldwide level. The interaction of angles formed among satellites and receivers serves as the fundamental basis upon which human civilization relies (Latibjonovich,

2022). Through this dynamic interaction, our technological gadgets serve as navigational tools, assisting us in traversing foreign landscapes and intricate urban environments by deciphering abstract signals and transforming them into practical routes (Suherman et al., 2018).

Trigonometry plays a crucial role in the orchestration of sound technology, including a wide range of auditory phenomena, from the resonance of the string of a guitar to the rotational motion of a fan. The concept of musical notes, which were previously perceived as intangible and elusive, can now be precisely defined by the repetitive patterns of sine waves (Prabowo et al., 2018). The emergence of the range of a bottom note and the resonant sound of a soprano's high voice pitch can be attributed to harmonics and frequency connections that are rooted in trigonometric concepts (N. Mosese & Ogbonnaya, 2021).

Tip

The origins of trigonometry may be attributed to ancient civilizations such as the Babylonians, the Greeks and Egyptians, who employed fundamental concepts of trigonometry for practical purposes including as survey and astronomy.

8.2. TRIGONOMETRY IN SCIENCE AND TECHNOLOGY

Trigonometry, intricately intertwined with mathematical expressions and functions, serves as a fundamental block in modern technologies. This relationship extends across diverse domains, where the principles of trigonometry facilitate advanced technological applications.

8.2.1. Engineering and Architecture

In engineering and architecture, trigonometry finds extensive use. The Pythagorean theorem $\left(c^2 = a^2 + b^2\right)$, a fundamental trigonometric identity, is employed to ascertain the lengths of sides in right triangles. Trigonometric functions such as sine, cosine, and tangent enable precise angle calculations, aiding in the design and construction of stable and geometrically accurate structures.

8.2.2. Physics and Mechanics

Trigonometry plays a pivotal role in the analysis of motion. It is instrumental in breaking down complex motions into simpler components. In projectile motion, for instance, trigonometric functions are utilized to determine launch angles θ and initial velocities v.

8.2.3. Navigation and Geodesy

Navigation systems, including GPS, rely extensively on trigonometry. The spherical trigonometry equations, governing triangles on the curved surface of the Earth, are employed to compute distances and directions accurately across the globe.

8.2.4. Electrical Engineering

Trigonometry is deeply rooted in electrical engineering, especially in alternating current (AC) circuit analysis. Concepts such as phase angles ϕ, impedance Z, and phasor diagrams are all based on trigonometric functions. Engineers use trigonometric identities to predict and optimize electrical circuit behavior.

8.2.5. Digital Signal Processing

Digital signal processing (DSP) utilizes trigonometric techniques for signal manipulation. Fourier analysis, the governing equation of DSP, involves decomposing complex signals into constituent sinusoidal components by using trigonometric functions.

8.2.6. Computer Graphics and Animation

Trigonometry forms the mathematical basis for computer graphics and animation. Algorithms for rendering 3D objects, simulating lighting and shadow effects, and animating objects heavily rely on trigonometric calculations (Andreescu & Feng, 2005; Van Brummelen, 2009; Hosch, 2010; Wildberger, 2015).

8.2.7. Robotics and Automation

In robotics and automation, trigonometry facilitates spatial calculations for robot motion. Kinematics, a vital area of robotics, employs trigonometric equations to enable precise robot movements and interactions with the environment (Van Ingen, 1914; Bandeira & de Souza, 2006).

8.2.8. Astronomy and Astrophysics

Trigonometry is a critical tool in astronomy and astrophysics. It aids in measuring distances between celestial objects, determining the positions of celestial bodies, and studying the geometrical properties of celestial phenomena (Wilson, 2008; Confrey et al., 2010).

Did You Know?

The Pythagorean Theorem, a renowned mathematical theory, establishes a fundamental connection between the sides of a triangle with a right angle, hence serving as a cornerstone in the field of trigonometry. The nomenclature of this concept derives from the renowned ancient Greek scientist Pythagoras.

8.3. COMPUTER GRAPHICS AND ANIMATION

In graphic arts, trigonometry assumes a prominent role by skillfully coordinating the array of visual phenomena that charm our perceptual faculties (Bhatti et al., 2016). Trigonometry plays a vital role in the creative and technical convergence by facilitating the creation of realistic characters in computer games and cinematic effects. It accomplishes this by translating mathematical principles into visually engaging elements (Weber, 2008; Cetin, 2015; Mosese, 2017; Weng et al., 2017) (Figure 8.1).

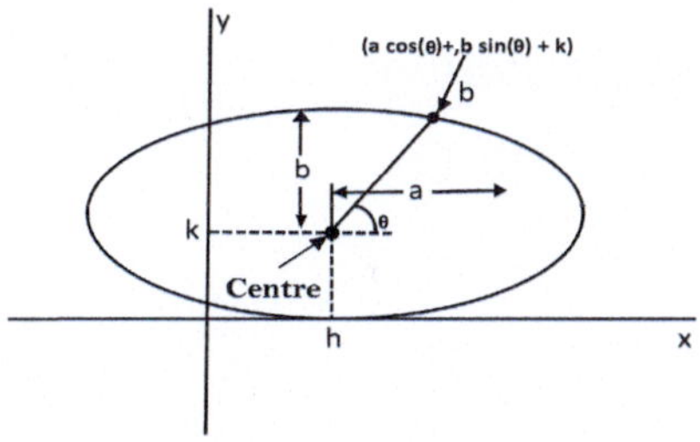

Figure 8.1. Computer graphics & trigonometric method.

Source: Frank, Creative Commons License.

8.3.1. Geometry and Transformation

Geometry assumes a fundamental role in computer graphics, while trigonometry serves as the conduit through which geometric modifications infuse the two-dimensional and three-dimensional realms. Translation, scaling, rotation, and shearing stand as the essential transformations employed by artists to mold virtual landscapes. At the core of these transformations lie the sine and cosine functions that translate angles into motion and give various forms to the artists' creations (Shaghaghian et al., 2022).

For instance, imagine a three-dimensional object undergoing rotational motion on a display surface. The rotation of the object is facilitated by an orchestrated sequence of cosine and sine waves, which dictate the movement of each vertex in precise accordance with trigonometric patterns (Brodetsky, 1924; Cundy, 1971; Lenahan, 2019; Mosese, 2017). The unit circle exhibits dynamic behavior when angles undergo alterations, which consequently results in a digital display (Yeshurun, 1982).

8.3.1.1. 2D Rotation Transformation

In computer graphics, a 2D rotation transformation is often expressed as:

$$\begin{bmatrix} cos(\theta) & -sin(\theta) \\ sin(\theta) & cos(\theta) \end{bmatrix} \begin{bmatrix} x \\ y \end{bmatrix}$$

where θ represents the angle of rotation, and (x, y) is a point in 2D space.

8.3.1.2. Scaling Transformation

Scaling a 2D object is represented as:

$$\begin{bmatrix} s_x & 0 \\ 0 & s_y \end{bmatrix} \begin{bmatrix} x \\ y \end{bmatrix}$$

Here, s_x and s_y are scaling factors along the x and y axes, respectively.

8.3.1.3. Translation Transformation

Translating a point (x, y) by (t_x, t_y) units is expressed as:

$$\begin{bmatrix} 1 & 0 & t_x \\ 0 & 1 & t_y \\ 0 & 0 & 1 \end{bmatrix} \begin{bmatrix} x \\ y \\ 1 \end{bmatrix}$$

This relation is often used to move objects in 2D space.

8.3.1.4. Homogeneous Coordinates

In computer graphics, homogeneous coordinates are used for transformations.

A point in 2D space $(x\ y)$ is represented as $(x, y, 1)$ in homogeneous coordinates.

8.3.1.5. Projection Transformation

For 3D to 2D projection, perspective projection is often represented using trigonometric functions:

$$\begin{bmatrix} f & 0 & 0 & 0 \\ 0 & f & 0 & 0 \\ 0 & 0 & A & B \\ 0 & 0 & -1 & 0 \end{bmatrix} \begin{bmatrix} X \\ Y \\ Z \\ 1 \end{bmatrix}$$

Here, f represents the focal length, and A and B are parameters for perspective correction.

8.3.1.6. Bézier Curves

Bézier curves used in computer graphics and animation are often defined using trigonometric functions and Bernstein basis polynomials. For instance, a quadratic Bézier curve is:

$$B(t) = (1-t)^2 P0 + 2(1-t)tP1 + t^2 P2$$

where t varies from 0 to 1, and $P0, P1, and\ P2$ are control points.

These mathematical expressions and transformations are fundamental in creating and manipulating objects and animations in computer graphics, illustrating the close relationship between trigonometry and digital visual representation.

8.3.2. The Unit Circle and Parametric Equations

The concept of unit circle establishes the fundamental principles behind parametric equations, which govern the motion of lines, points, and objects. Artists use the unit circle as a tool to delineate trajectories and establish curves (Kabaca & Aktumen, 2010).

Envision a rocket ascending through the atmosphere while emitting a luminous trail of ignited particles. This trajectory is governed by parametric equations, fundamental characteristics of the unit circle (Jones et al.,1989). Through the manipulation of trigonometric functions, designers have the ability to generate complex flight routes, replicate the movement of fluids, and create dynamic combinations of light and shadow (Cruz-Barroso et al., 2007).

Here are some mathematical expressions related to the Unit Circle and Parametric Equations in the context of trigonometry, particularly relevant in computer graphics and animation:

Unit Circle Equation: The unit circle, centered at the origin (0,0) with a radius of 1, is defined by the equation:

$$x^2 + y^2 = 1$$

This equation represents all points (x, y) on the unit circle.

- **Trigonometric Definitions:** The unit circle also serves to define trigonometric functions. For a point (x, y) on the unit circle corresponding to an angle θ:

$$\cos(\theta) = x$$

$$\sin(\theta) = y$$

$$\tan(\theta) = \frac{y}{x}$$

- **Parametric Equations:** Parametric equations describe the coordinates (x, y) of a point in terms of a third parameter, typically denoted as 't.' In computer graphics and animation, parametric equations are used to create curves and animations.
- **Parametric Equations for a Circle:** To draw a circle centered at (a, (b) with radius 'r,' use parametric equations:

$$x(t) = a + r\cos(t)$$

$$y(t) = b + r\sin(t)$$

Here, 't' is the parameter that varies, and as 't' varies from 0 to 2π, it traces out a complete circle.

- **Parametric Equations for Animation:** In computer animation, parametric equations can describe the motion of objects over time. For instance, to model the motion of an object along a curved path, one may use parametric equations for position (x, y) as functions of time 't.'
- **Parametric Equations for Bézier Curves:** Bézier curves are fundamental in computer graphics. They can be defined using parametric equations. For a quadratic Bézier curve with control points P0, P1, and P2, the parametric equations for (x, y) are:

$$x(t) = (1-t)^2 \times P0_x + 2(1-t)t \times P1_x + t^2 \times P2_x$$

$$y(t) = (1-t)^2 \times P0_y + 2(1-t)t \times P1_y + t^2 \times P2_y$$

Here, 't' varies from 0 to 1, and as it varies, it traces the Bézier curve.

These mathematical expressions, involving the unit circle and parametric equations, are foundational in computer graphics and animation. They enable the creation of visually appealing and dynamic animations, providing a mathematical basis for modeling and rendering objects and their movements.

8.3.3. Rotations in 3D Space

Within the domain of three-dimensional graphics, trigonometry assumes a 3-D manifestation, infusing virtual sceneries with depth and perspective. Rotations progressively increase in complexity as angles expand over many axes. Euler angles, as well as quaternion rotations, which are mathematical representations of spatial orientations, enable artists to manipulate gravity and induce rotational motion in galaxies. These transformations are facilitated by trigonometric functions, which play a crucial role in the calculation of these rotations (Kabaca & Aktumen, 2010).

An illustration of this involves the utilization of animation to depict a three-dimensional camera executing a circular trajectory around an item. By employing complex trigonometric computations, the camera executes a sweeping motion, revealing the picture from several perspectives. This immersive experience arises from the integration of quantitative precision with artistic vision (Jones et al., 1989).

8.3.4. Shaders and Lighting

Trigonometry is also utilized in techniques of shading and lighting, contributing to the creation of authenticity and perspective in computer-generated sceneries. Artists employ mathematical calculations to recreate the complex interactions of shadows and light by determining the angles between illumination, surfaces, and the spectator (Ellaithy et al., 2017). The calculations, which are based on trigonometric principles, shape the physical features of objects and elicit emotional responses by manipulating the interaction between light and shadow (Osudin, 2022).

For example, in a simulation, the dynamic interaction between sunshine and shadow, as well as the reflections on the surface of a lake, may be understood as intricate mathematical compositions orchestrated by the rules of trigonometry (Nam et al., 2007).

8.3.5. Creating Visual Illusions

Trigonometry demonstrates its use within the domain of visual illusions, wherein artists employ the manipulation of angles to engender perceptual phenomena that elicit profound cognitive dissonance. The creation of anamorphic art, which involves the production of distorted pictures that look cohesive when viewed from a certain vantage point, is dependent on the precise application of trigonometric calculations to achieve the desired distortions. The visual appeal of optical illusions, characterized by the perception of skewed lines and shifting objects, is rooted in the mathematical principles of trigonometric relationships (Sparks, 2023).

Within the field of graphic design and animation, trigonometry orchestrates a fusion of mathematical principles and artistic expression (Landwehr, 2022). Trigonometry plays a pivotal role in translating complex mathematical concepts into visual narratives, ranging from algorithms that replicate rainfall on a piece of glass to simulations that transport us into imaginative universes. With the continuous advancement of technology and the persistent exploration of creative limits by artists, trigonometry continues to serve as a constant source of inspiration, fueling new realms of imagination and potential (Jackson, 2021). Here are some mathematical expressions that illustrate how these illusions can be achieved:

- **Rotating Objects:** To create the illusion of an object rotating smoothly, you can use the parametric equations for circular motion:

$$x(t) = r \cdot \cos(\omega t)$$

$$y(t) = r \cdot \sin(\omega t)$$

Here, r represents the radius of the rotation, ù is the angular velocity, and t denotes time. By varying ù, one can control the speed of rotation.

- **Waving Flags:** Simulating the waving of a flag or a banner can be achieved through sinusoidal motion. For instance, to depict a flag flapping in the wind:

$$y(t) = A \cdot \sin(\omega t)$$

Here, A represents the amplitude of the wave, and ù determines the frequency of the waving of flag.

- **Pulsating Effects:** To create pulsating effects, trigonometric functions can be used. For instance, a pulsating light effect can be represented as:

$$I(t) = I_0 + A \cdot \sin(\omega t)$$

Here, $I(t)$ represents the intensity of the light at time t, I_0 is the base intensity, A is the amplitude of variation, and ù controls the frequency of pulsation.

- **3D Animation:** In 3D animation, trigonometric functions are employed to create smooth camera movements. For example, to control the camera's position in a circular orbit around an object:

$$x(t) = r \cdot \cos(\omega t)$$

$$y(t) = r \cdot \sin(\omega t)$$

$$z(t) = z_0$$

Here, r is the orbit radius, ù sets the angular velocity, and z_0 keeps the camera at a constant height.

Morphing Shapes: Trigonometric functions can be used for shape morphing. For instance, to morph a shape from one to another:

$$x(t) = (1-t) \cdot x_1 + t \cdot x_2$$

$$y(t) = (1-t) \cdot y_1 + t \cdot y_2$$

Here, (x_1, y_1) and (x_2, y_2) represent the initial and final positions of the shape, and (t) varies from 0 to 1 to control the morphing.

These mathematical expressions, often implemented through programming languages and tools in computer graphics and animation software, enable artists and developers to create captivating visual illusions and animations by harnessing the power of trigonometry.

Remember

The utilization of a unit circle is a highly effective and influential method within the field of trigonometry. It has a diameter of 1 unit, while its angular measurements are expressed in radians. Trigonometric functions such as sine and cosine may be formally defined by utilizing the location of points situated on the unit circle.

8.4. ENGINEERING AND ARCHITECTURE

This section delves into how trigonometry underpins the structural integrity of buildings, bridges, and infrastructure, ensuring their stability and safety. This section also uncovers how engineers employ trigonometry for complex tasks like surveying vast landscapes, designing efficient transportation systems, and harnessing renewable energy sources. In the realm of architecture, trigonometry guides the creation of aesthetically pleasing and functional structures, facilitating the translation of innovative design concepts into tangible constructions.

The following sections reveal trigonometry as an indispensable tool for engineers and architects, enabling them to address a multitude of practical challenges. Understanding the role of trigonometry in these fields offers insights into the essential synergy between mathematics and the built environment, emphasizing the profound impact of this mathematical discipline on the physical world (Blanquer et al., 2017) (Figure 8.2).

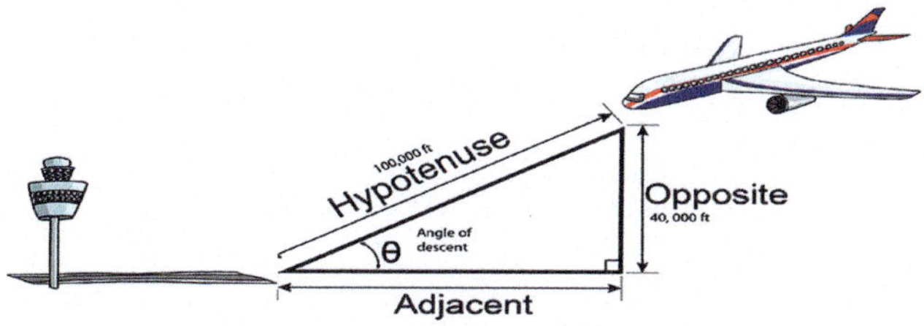

Figure 8.2. Trigonometry in flight engineering.

Source: Martin, Creative Commons License.

8.4.1. Structural Analysis and Design

Trigonometry plays a fundamental role in the field of structural engineering, providing the necessary basis for the examination and development of various architectural structures such as buildings, roads, and other forms of infrastructure. Trigonometric relationships are utilized by engineers in order to compute the forces and stresses that structures are subjected to. Trigonometry plays a crucial role in several aspects of structural engineering, such as estimating the angles of beam intersections and calculating load distributions within complex frameworks. It helps ensure that buildings possess the necessary strength and stability to withstand both natural forces and human activities.

The creation of a suspended bridge is a remarkable feat of engineering, as it is designed to span great distances. Trigonometry plays a crucial role in the analysis of angles and forces that govern cable tensions, assuring the maintenance of balance and safety. By employing trigonometric computations, engineers achieve a meticulous equilibrium between aesthetics and practicality, ensuring the longevity and inspirational impact of these remarkable architectural marvels (Melchels et al., 2010).

- **Example:** Calculation of Bending Moment in a Simply Supported Beam

Consider a simply supported beam with a uniform load q along its length L. To calculate the bending moment $M(x)$ at any point x along the beam, the following mathematical expression can be used:

$$M(x) = \frac{q}{2}\left(L^2 - x^2\right)$$

This mathematical expression allows engineers to determine the bending moment at any point along the beam, which is crucial for designing structural elements that can safely withstand the applied loads.

8.4.2. Surveying and Land Measurement

The process of surveying, which is essential for land measuring and mapping, is significantly reliant on the concepts of trigonometry. Surveyors employ trigonometry to precisely measure angles and distances while undertaking tasks such as delineating property borders, charting topographical features, or devising urban plans. Surveyors utilize trigonometric functions to triangulate locations, determine elevations, and generate maps that facilitate planning for cities and resource management (Ni et al., 2015).

Envision surveyors engaged in the process of charting the complex topographical features of an urban environment. Accurate depictions of the topography are created by employing trigonometric concepts to measure angles among landmarks and calculate distances. These cartographic representations, founded on rigorous mathematical accuracy, serve as essential tools for designers, mathematicians, and policymakers in the process of designing cities that achieve a harmonious integration with the natural surroundings (Loudon, 1845) (Figure 8.3).

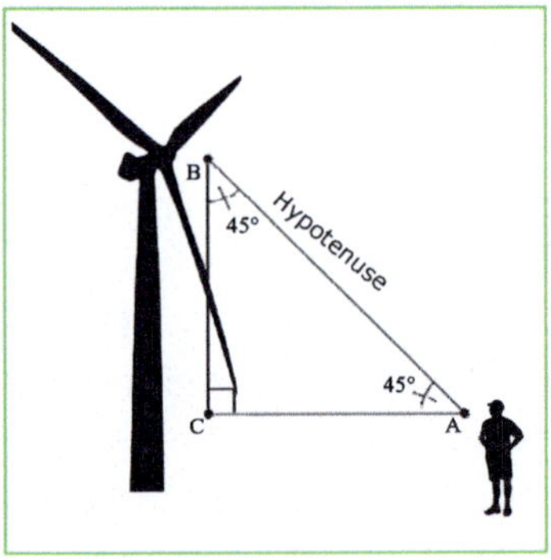

Figure 8.3. Trigonometry in engineering fields.

Source: Martin, Creative Commons License.

8.4.3. Architectural Design and Aesthetics

Trigonometry's reach expands beyond its role in ensuring structural stability to encompass a substantial influence on the aesthetic dimensions of architecture. Architects leverage trigonometric principles to determine angles that maximize the utilization of natural light, minimize energy consumption, and create visually pleasing designs. Through the calculation of angles for optimal sunlight penetration, architects craft energy-efficient structures that adeptly harness natural illumination (Ziatdinov et al., 2012).

The phenomenon of sunlight filtering through the openings of a church is an instance of this. Trigonometry plays a crucial role in determining the optimal positioning of windows, enabling the entry of sunshine to permeate stained glass designs, thus enhancing the illumination of sacred areas. Trigonometry's impact encompasses a wide range of phenomena, spanning from the intricate spirals observed in seashells that adhere to the Fibonacci sequence to the application of the golden ratio in the construction of renowned architectural masterpieces (Liu et al., 2015).

Example: The Golden Ratio in Architectural Proportions

The Golden Ratio ϕ is a mathematical constant approximately equal to 1.6180339887 and is often found in architectural design. It is defined by the relationship:

$$\phi = \frac{a+b}{a} = \frac{a}{b}$$

Where:

a is the longer side of the object;

b is the shorter side of the object.

In architecture, the Golden Ratio is employed to create aesthetically pleasing proportions in structures. For instance, the ratio of the height of a building to the width of its base or the dimensions of a room can often be close to the Golden Ratio. This mathematical harmony is believed to enhance the visual appeal of architectural designs, contributing to their aesthetic quality and balance.

8.4.4. Mechanical Engineering and Kinematics

Trigonometry is used in the field of mechanical design, where its concepts are utilized to guide the motion of equipment and mechanisms. Trigonometric equations are utilized by engineers in the design of gears, connections, and robotic arms, with the aim of achieving motion that is both smooth and accurate. The field of kinematics, which pertains to the analysis of motion, heavily relies on the principles of trigonometry in order to determine angles, velocities, and accelerations that are fundamental to the operation of mechanical systems (Petrescu et al., 2017).

Consider the scenario of an automated arm engaged in the task of assembling complex components on a production line. The precise movements of the arm, which are determined by trigonometric equations, serve as a notable illustration of the level of accuracy that trigonometry contributes to mechanical procedures. Trigonometry plays a pivotal role in shaping the mechanical world across several domains, ranging from industrial automation to aerospace design (Trinh et al., 2015).

Example: The Equation of Uniformly Accelerated Motion

In mechanical engineering and kinematics, one fundamental equation describing the motion of an object under constant acceleration is the equation of motion:

$$s = ut + \frac{1}{2}at^2$$

Where:

s represents the displacement of the object from its initial position.

u denotes the initial velocity of the object.

a is the constant acceleration.

t signifies the time elapsed since the start of the motion.

This equation quantitatively describes how an object's position s changes over time t when subjected to a constant acceleration a, considering its initial velocity u. It finds extensive application in mechanical engineering for predicting the positions and motions of objects in various scenarios, from simple projectile motion to the movement of vehicles and machinery.

8.4.5. Geometric Constructions and Artistry

The utilization of trigonometry in architecture and engineering goes beyond mere mathematical computations, as it exhibits an interconnection with the dimensions of design. Trigonometry serves as a guiding principle for the implementation of geometric constructs, which in turn provides the foundation for the elaborate patterns, decorations, and aesthetics that decorate various buildings. The combination of ratios based on trigonometry and architectural ingenuity gives rise to the rhythmic recurrence of arches, the perfect alignment of facades, and the spiraling staircases (Brodetsky, 1919).

The elaborate patterns of muqarnas in Islamic architecture exemplify the application of trigonometric ideas through the use of interconnecting geometric forms. The presence of these patterns, which can be observed in mosques and palaces, serves as evidence of the seamless integration of mathematics and artistic expression. In this context, the meticulous calculations of trigonometry contribute to the enhancement of the visual appeal found in architectural marvels.

Within the realms of architecture and engineering, trigonometry extends beyond its mathematical foundations to emerge as the fundamental basis for design, accuracy, and visual appeal. The quiet force referred to in this statement is a crucial factor that provides engineers with the ability to connect the conceptualization process with the actual building phase. This force grants them the capability to design and construct structures that possess the durability to endure the test of time and various environmental conditions. Trigonometry serves as a tribute to humanity's imagination and ability, influencing the physical world via the construction of soaring buildings, bridges, rigorous surveying techniques, and subtle aesthetic embellishments. Its impact on the landscapes of the future is undeniable, leaving a lasting impression (Reid et al., 2009).

Example: Consider the construction of an Islamic eight-pointed star, a fundamental motif in Islamic art. This star, known as an "Octagram," is created by intersecting two squares, one rotated by 45 degrees relative to the other.

Mathematically, this can be represented by the equations for the coordinates of the star's points:

For the outer square with side length a, the coordinates of its vertices are:

$$A(0,a), B(a,0), C(0,-a), D(-a,0).$$

For the inner square with side length a and rotated by 45 degrees, the coordinates of its vertices are:

$$E\left(a/\sqrt{2}, a/\sqrt{2}\right), F\left(a/\sqrt{2}, -a/\sqrt{2}\right), G\left(-a/\sqrt{2}, -a/\sqrt{2}\right), H\left(-a/\sqrt{2}, a/\sqrt{2}\right).$$

The Octagram's points are formed by the intersection of lines connecting the midpoints of opposite sides of the outer square with the midpoints of adjacent sides of the inner square.

I is the midpoint of AD, and J is the midpoint of BC.

The coordinates of the Octagram's points are:

$$I(0, a/2)\, and\, J(a/2, 0)$$

These mathematical expressions guide the precise construction of the Octagram, showcasing the seamless integration of mathematics and artistic creativity in Islamic geometric patterns.

Remember

Trigonometric identities are mathematical equations that are universally valid for all angles. The Pythagorean identity is often recognized as the most prominent. $\sin^2(\theta) + \cos^2(\theta) = 1$.

8.5. NAVIGATION & GPS TECHNOLOGY

The field of navigation has experienced a significant transformation with the introduction of Global Positioning System (GPS) technology, which has revolutionized the process of determining one's location and navigating around the expansive terrains of the globe. Trigonometry, a perennial mathematical field, serves as the fundamental basis for this transformative process, enabling precise and efficient navigation of the contemporary environment (Huang et al., 2020).

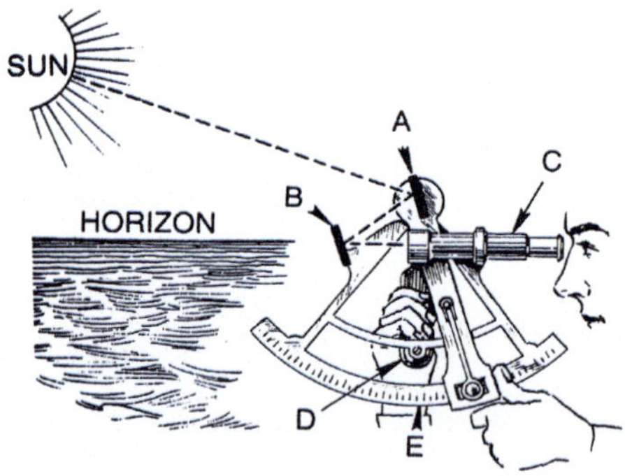

Figure 8.4. Application of trigonometry in real life.

Source: Frank, Creative Commons License.

Trigonometry has undergone a significant transformation over time, transitioning from the use of sextants by ancient seafarers to the reliance on GPS-equipped gadgets by contemporary travelers. This evolution has enabled trigonometry to play a crucial role in navigating both unexplored regions and densely populated urban areas (Vaitheeswaran et al., 2012) (Figure 8.4).

8.5.1. Triangulation and Navigational Bearings

The fundamental principle behind navigation is triangulation, a technique that originates from the field of trigonometry. Navigators are able to determine their precise location by employing the method of calculating angles between points of reference and a distant destination. For ages, seafarers have utilized the notion of utilizing the positions of the stars and monuments to infer their location. The utilization of trigonometric relationships facilitated the computation of distances and orientations, enabling individuals to navigate across the seas with remarkable accuracy (Rolls, 2021).

The Polynesian voyagers are a subject of interest due to their remarkable ability to navigate the expansive Pacific Ocean, only relying on celestial bodies and stars as navigational aids. The navigators of that era heavily depended on trigonometric calculations in order to ascertain their precise latitude and longitude. This was a sophisticated process of meticulously coordinating various angles and locations, which extended across vast distances (Lee et al.,2014).

Triangulation Example: Suppose we have three fixed points A, B, and C, with known coordinates in a two-dimensional plane. To determine the location of an unknown point X, we can use the Law of Sines:

$$\frac{\sin(\angle AXB)}{AB} = \frac{\sin(\angle BXC)}{BC} = \frac{\sin(\angle CXA)}{AC}$$

Where:

$\angle AXB$ is the angle at point X between lines AX and BX.

$\angle BXC$ is the angle at point X between lines BX and CX.

$\angle CXA$ is the angle at point X between lines CX and AX.

AB, BC, and AC are the known distances between points A and B, B and C, and C and A, respectively.

By measuring these angles and knowing the distances, we can use trigonometric calculations to determine the coordinates of point X, enabling precise triangulation.

8.5.2. GPS and Satellite Triangulation

Trigonometry plays a vital role in the GPS, or the Global Positioning System, in contemporary times. GPS technology is based on the concept of triangulation, which utilizes a collection of satellite in Earth's orbit to accurately ascertain specific positions on the Earth's surface. GPS systems are capable of determining the distances between a receiver and several satellites by employing calculations based on the time it takes for signals to make their way from these satellites to the receiver. The aforementioned distances, when coupled with the established positions of the satellites, enable the precise determination of locations by equipment with remarkable precision (Hwang & Hwang, 2002).

Consider a hypothetical scenario in which an individual, acting as a traveler, is effectively maneuvering around a densely populated urban area with the assistance of a GPS gadget. The system utilizes the concepts of trigonometry to combine signals from many satellites and accurately calculate coordinates. The integration of mathematics into technology serves as a testament to its ability to navigate the intricate nature of urban environments and foreign topographies (Strauß, 2005).

- **The Distance Equation:** In GPS and satellite triangulation, the distance d between a receiver and a satellite can be calculated using the speed of light c, the time it takes for a signal to travel from the satellite to the receiver t, and the equation:

$$d = \frac{c \cdot t}{2}$$

Here, d represents the distance between the receiver and the satellite, c is the speed of light (approximately $299,792,458$ meters per second), and t is the time taken for the signal to travel from the satellite to the receiver. This equation forms the basis for determining the distance to multiple satellites, allowing GPS receivers to triangulate their positions accurately in three dimensions.

8.5.3. Trilateration and Precision

Trilateration, a variant of the triangulation method, assumes a crucial part in the functioning of GPS technology. The process entails the quantification of spatial distances as opposed to angular measurements. Through the utilization of intersecting circles, or in three-dimensional space, spheres centered around established reference points, navigators are able to ascertain the precise coordinates of an unfamiliar spot. This methodology, facilitated by the utilization of trigonometric computations, guarantees that GPS receivers are capable of determining precise locations even under complex and demanding conditions (Yang et al., 2017).

Let us contemplate an investigation and rescue endeavor conducted within a challenging and treacherous natural environment. GPS devices employ trilateration techniques to aid rescuers in the process of locating persons who are reported missing. This is achieved by analyzing the signal strength received from various satellites. These gadgets facilitate complex trigonometric computations in real-time, enabling rescuers to cross hazardous terrains and deliver prompt aid (Liang & Krause, 2014).

8.5.4. Navigation in Urban Environments

GPS technology has distinct problems in metropolitan contexts characterized by convoluted streets and obstructed views of the sky. The occurrence, sometimes referred to as the "urban canyon" phenomenon, arises when the presence of towering architectural structures obstructs the effective reception of signals from satellites. In order to address this challenge, GPS receivers depend on the use of both straight and reflected signals, employing trigonometric principles to compute distances by analyzing the time it takes for signals to travel.

Consider a motorist maneuvering in a highly populated urban environment. The utilization of GPS technology, supported by trigonometric computations, enables the determination of the precise position of a vehicle, even in scenarios where tall buildings obstruct the road and produce shadows. The seamless integration of technology and mathematics allows precise navigation across the intricate urban landscape (Bedogni et al., 2015).

The ongoing advancement of technology will inevitably lead to the evolution of trigonometry's significance in the fields of navigation and GPS technology. (Johnson & Walker) The use of Augmented Reality, machine learning, and sophisticated techniques will enhance its functionalities, facilitating more accurate and smooth navigation encounters (Trinh et al., 2015).

Trigonometry has had a transformative journey over the course of human history, transitioning from its initial use in aiding sailors in celestial navigation to its current significance in navigating the vast digital landscape of our linked global society (Suherman et al., 2018). The enduring principles underlying GPS technology have been effectively utilized to change global navigation. From ancient seafarers to contemporary individuals equipped with smartphones, trigonometry continues to serve as a steadfast navigational tool, assisting us in navigating the intricate dimensions of space and time (Li et al., 2018).

Certainly, a fundamental mathematical expression related to navigation in urban environments, specifically for estimating the time of arrival (ETA) of a vehicle or pedestrian, involves:

ETA Calculation:

$$ETA = \frac{d}{v}$$

In this expression, ETA represents the estimated time of arrival, d denotes the remaining distance to the destination, and v signifies the velocity or speed at which the vehicle or pedestrian is traveling. This equation is used in navigation systems to provide real-time estimates of when a vehicle or person will reach their destination based on their current speed and the remaining distance to travel.

Remember

The sine, cosine, and tangent functions are the three fundamental trigonometric ratios. These ratios elucidate the interconnections between the various sides in a right triangular and the corresponding angles.

8.6. TRIGONOMETRY IN SOUND & MUSIC TECHNOLOGY

Within the realm of symphonic music and sound, where melodies move, and harmonics intertwine, trigonometry appears as an imperceptible conductor, skillfully coordinating the complicated interaction of frequencies and waveforms. This field of study serves as an intermediary between scientific abstraction and creative expression, enriching our comprehension of the propagation of sound waves, resonance in musical instruments, and the creation of musical compositions. Trigonometry significantly influences the audio landscapes that surround us, ranging from the melodies produced by a violin to the powerful rhythms from a bass drum (Giordano & Wanderley, 2013).

8.6.1. Sine Waves and the Foundation of Sound

The sine waves, a fundamental idea that encapsulates the fundamentals of periodic oscillations, are at the core of trigonometry's impact on sound technology. Sine waves exhibit a distinct set of properties, including their continuous and regular oscillation pattern, which represents the epitome of undistorted vibrational motion. Trigonometric functions, such as the sine and cosine, play a pivotal role in generating these waves, elucidating the associations among the three variables: frequency, amplitude, and time (Figure 8.5).

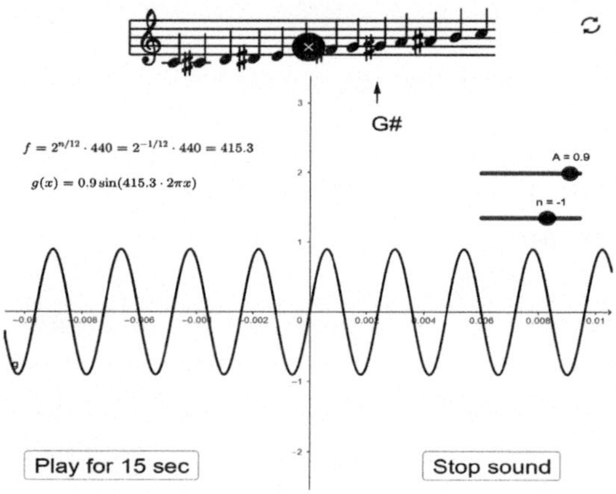

Figure 8.5. Trigonometry in music.

Source: Brian, Creative Commons License.

Envision the gentle resonance produced by a tuning fork. The hymn may be described as a wave that resembles a sine in its most basic manifestation, characterized by unique frequency and tone. By utilizing principles from trigonometry, it becomes possible to deconstruct sophisticated auditory stimuli into their constituent sine wave elements, therefore revealing the intricate melodic structures that comprise a symphony (Wilhelm, 2002).

One fundamental mathematical expression related to trigonometry in sine waves and the foundation of sound is as follows.

- **The Wave Equation**: In the context of sound and sine waves, the fundamental wave equation is represented as:

$$y(t) = A\sin(2\pi ft + \phi)$$

Here, $y(t)$ represents the instantaneous displacement of the wave at time t, A is the amplitude of the wave (its maximum displacement from the equilibrium position), f is the frequency (the number of oscillations per unit time), t is time, and ϕ is the phase angle.

This equation encapsulates the essence of sound waves, illustrating how sound pressure variations are modeled as sinusoidal oscillations. It is foundational in understanding the mathematical underpinnings of sound, where trigonometric functions like sine characterize the waveforms that make up sound waves.

8.6.2. Harmonics and Fourier Analysis

Within the domain of music and sound, musical instruments do not generate singular sine waves in isolation. In contrast, these acoustic phenomena give rise to intricate amalgamations of frequencies referred to as harmonics and overtones. The presence of harmonics contributes to the timbre and distinctive qualities of the sound produced by each individual instrument. Trigonometry facilitates the examination of harmonic structures by means of the Fourier transform, a mathematical technique that deconstructs intricate waveforms into their individual frequencies (Lostanlen et al., 2019).

Let us contemplate a guitar string that has been plucked. When undergoing vibration, the object generates a diverse array of frequencies that collectively form its distinct auditory profile. By use of Fourier analysis, the discipline of trigonometry reveals the existence and magnitude of individual harmonics, offering a valuable understanding of the timbre of a guitar. This knowledge empowers sound engineers to manipulate the audible characteristics of recordings and live performances, thereby shaping the overall auditory identity (Cruz et al., 2020).

- **The Fourier Series Expression:** In the context of harmonics and Fourier analysis, the Fourier series representation of a periodic function $f(t)$ is given by:

$$f(t) = \frac{a_0}{2} + \sum_{n=1}^{\infty}\left[a_n \cos\left(\frac{2\pi n}{T}t\right) + b_n \sin\left(\frac{2\pi n}{T}t\right)\right]$$

In this expression, $f(t)$ represents a periodic function with a period

T. The terms a_0, a_n, and b_n are coefficients that can be calculated using integrals and represent the amplitudes of the different harmonic components. The expression essentially decomposes the periodic function into a sum of sinusoidal functions (cosines and sines) with different frequencies (harmonics), providing a powerful tool for analyzing and synthesizing complex waveforms in various fields, including signal processing, music, and engineering.

8.6.3. Synthesis and Sound Design

The field of sound generation and design is significantly impacted by the principles of trigonometry since artists and sound engineers utilize these principles to create and shape intricate aural landscapes. Oscillators, which serve as fundamental components in the synthesis of sounds, utilize trigonometric functions to produce waveforms that imitate conventional musical instruments or explore novel auditory domains.

8.6.4. Modulation and Effects

Following is a fundamental mathematical expression related to trigonometry in modulation and effects, specifically in the context of amplitude modulation (AM).

- **Amplitude Modulation (AM) Expression:** In AM, a carrier signal (t) is modulated by a message signal $M(t)$ to produce a modulated signal $S(t)$. The mathematical expression for the AM signal is given as:

$$S(t)=A_c \cdot \cos(2\pi f_c t) \cdot [1+k_a \cdot M(t)$$

This expression illustrates how the amplitude of the carrier signal is modulated by the message signal, producing sidebands around the carrier frequency.

8.6.5. Digital Signal Processing

In the contemporary era of technology, the field of music has witnessed significant advancements, particularly in the realm of DSP. This branch of study involves the utilization of algorithms to modify sound waves in order to get certain auditory outcomes. The utilization of trigonometry is readily apparent in several procedures, including filtering, equalizing, and pitch shifting. By employing meticulous calculations based on trigonometric principles, engineers possess the ability to manipulate sound with a high degree of precision (Vaseghi, 2007).

Consider the scenario of a disc jockey engaging in the process of remixing a musical recording via digital technology. Every modification, whether it is the isolation of a vocal file or the alteration of bass frequency, necessitates complex trigonometric computations. These calculations, performed within a millisecond timeframe, interlace the foundation of the remix, showcasing the instantaneous influence of trigonometry inside contemporary music production.

The Integration of trigonometry with music and sound technology extends beyond the confines of mathematical abstract and aesthetic expression. The auditory phenomenon of music is elevated to extraordinary levels through the orchestration of sine waves, the intricate interplay of the harmonics, and the refined technique of modulation. Trigonometry plays a significant role in the creation of music, whether through the melodic sounds produced by a guitar or the sophisticated digital manipulations of electronic music. It serves as an imperceptible composer, enhancing our auditory environments and facilitating the harmonious integration of mathematical principles with musical compositions (Park, 2009).

8.7. TRIGONOMETRY IN ELECTRONICS AND SIGNAL PROCESSING (FIGURE 8.6)

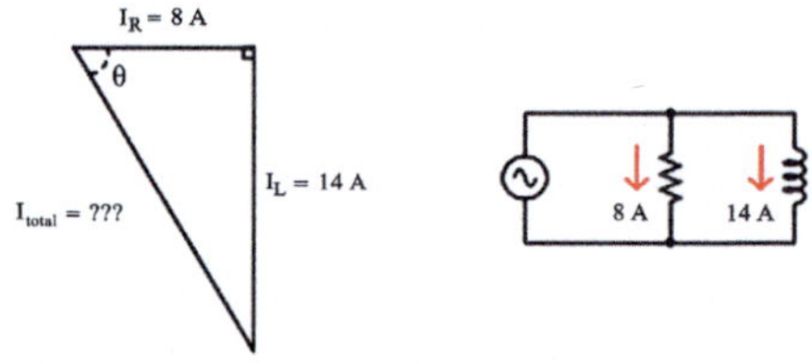

Figure 8.6. Trigonometry for AC circuits.

Source: Brian, Creative Commons License.

8.7.1. Waveforms and Frequency Analysis

Waveforms are fundamental components of electrical signals since they depict the dynamic variations in current or voltage over a certain period.

Trigonometry assumes a crucial role in comprehending and examining these waveforms. Engineers are able to obtain valuable insights into frequency, harmonics, and modulation by decomposing complicated signals into their individual sinusoidal components.

Let us contemplate the scenario of an audio engineer engaged in the process of manipulating and refining a musical composition. Trigonometry enables the analysis of sound waves by decomposing them into their constituent basic frequencies, including harmonics, enabling precise modifications that enhance auditory perception. Trigonometry enables engineers to manipulate sound with mathematical accuracy, whether it involves isolating the timbre of an instrument or rectifying phase imbalances (Bayeh & Mastorakis, 2013).

- **Example:** The Discrete Fourier Transform (DFT) Expression

In the realm of frequency analysis, the DFT is a crucial tool for converting a discrete-time signal $x[n]$ into its frequency-domain representation $X[k]$. The DFT equation for a discrete signal of length N is given by:

$$\left[X[k]=\sum_{n=0}^{N-1} x[n]\cdot e^{-j\frac{2\pi}{N}kn}\right]$$

Here, $X[k]$ represents the complex amplitude of the k-the frequency component, $x[n]$ is the discrete-time signal, N is the number of samples in the signal, and j denotes the imaginary unit.

This expression illustrates how trigonometric functions (through the complex exponential $e^{-j\frac{2\pi}{N}kn}$) are employed in frequency analysis to decompose a discrete signal into its constituent frequency components. It's an essential tool in fields like DSP and telecommunications, enabling the analysis and manipulation of signals in the frequency domain.

8.7.2. Frequency Filters and Equalization

Within the domain of signal processing, frequency filters play a crucial role as necessary instruments that effectively segregate desirable frequencies from undesired noise or interference. Trigonometric functions play a crucial role in the design and execution of these filters, allowing engineers to precisely target desired frequency ranges with a high degree of accuracy.

Consider a hypothetical scenario in which a noise-canceling headset is utilized to effectively suppress the auditory disturbances emanating from a densely populated subway environment. Trigonometry plays a crucial role in the development of algorithms that are designed to detect and mitigate undesirable frequencies, enabling consumers to have a tranquil aural environment while enjoying their music (Merched, 2006).

8.7.3. Modulation and Carrier Signals

Modulation, is the process of incorporating data into carrier messages. Trigonometric relationships are utilized in the methods of AM (Bandeira & de Souza, 2006) as well as frequency modulation, or FM, to determine the modulation of the carrier signal's intensity or frequency by the information signal, respectively.

Let us contemplate the procedure involved in the transmission of a radio signal. Trigonometry plays a crucial role in the process of combining the information-laden signal with the carrier signal, enabling individuals to effectively tune in to specific stations and experience audio broadcasts (Moreolo et al., 2009).

- **Example:** Amplitude Modulation (AM) Expression

In AM, a carrier signal $C(t)$ is modulated by a message signal $M(t)$ to produce an AM signal $S(t)$ using the following mathematical expression:

$$S(t) = A_c \cdot \cos(2\pi f_c t) \cdot [1 + k_a \cdot M(t)]$$

Here;

$S(t)$ represents the AM signal.

(A_c) is the amplitude of the carrier signal.

(f_c) is the frequency of the carrier signal.

(k_a) is the amplitude sensitivity factor, representing how much the carrier signal's amplitude varies with the message signal.

$M(t)$ represents the message signal, often a baseband signal.

This expression illustrates how the message signal is encoded onto the carrier signal by varying its amplitude. The trigonometric function ðos$(2\ f_c t)$ represents the carrier wave, and the modulation term $1 + k_a \cdot M(t)$ modulates the carrier's amplitude in accordance with the message signal $M(t)$. AM is a fundamental modulation technique used in radio broadcasting and communication systems.

8.7.4. Fourier Transform and Signal Representation

The Fourier Transform, a very influential mathematical technique based on trigonometry, facilitates the conversion between the temporal domain and the spectral domain for engineers. The process involves converting a signal through its waveform depiction to its frequency depiction, therefore revealing its intricate harmonic composition.

Consider an oscilloscope exhibiting the visualization of an intricate electrical waveform. By employing the Fourier Transform, the given waveform may be subjected to a transformation process, resulting in a spectrum that effectively illustrates the many frequencies that comprise it.

8.7.5. Digital Signal Processing (DSP)

In the contemporary era of digital technology, DSP, or digital signal processing, assumes a prominent role, wherein algorithms are employed to manipulate signals inside the digital realm. Trigonometry plays a crucial role in facilitating procedures such as Rapid Fourier Transformations (RFT), which enhance the efficiency of signal processing through the utilization of trigonometric symmetries and connections.

Consider a hypothetical scenario in which a smartphone utilizes voice analysis technology to effectively eliminate noise from the background during a telephone conversation. The algorithms employed utilize trigonometric computations to separate the user's speech from the ambient auditory environment, hence guaranteeing effective communication (Zhao et al., 2017).

8.7.6. Data Compression and Transmission

The field of trigonometry plays a significant part in reducing data size and transmission, which are crucial aspects of digital communication systems. Trigonometry is employed in compression algorithms to enhance data representation efficiency, resulting in reduced file sizes without compromising crucial information. Trigonometric functions serve as the foundation for modulation techniques employed in transmission, facilitating the encoding of data onto carrier signals. This enables the achievement of efficient and robust communication, minimizing the occurrence of errors.

Compression techniques that utilize trigonometry effectively decrease the amount of data needed to encode a picture, hence enabling efficient transmission and rapid display on a screen (Kekre & Solanki, 1978).

- **Example:** The Shannon-Hartley Theorem

In the context of data transmission, the Shannon-Hartley theorem provides an expression that relates the maximum data transmission rate (C) in bits per second over a channel with limited bandwidth and the signal-to-noise ratio (SNR) of the channel:

$$C = B \cdot \log_2 (1 + \text{SNR})$$

In this expression, C represents the maximum achievable data transmission rate (channel capacity) in bits per second, B is the bandwidth of the channel in hertz, and "SNR" is the signal-to-noise ratio, which quantifies the quality of the signal in the channel.

This theorem is fundamental in the design of data compression and transmission systems. It helps determine the maximum data rate that can be reliably transmitted over a noisy channel, taking into account the available bandwidth and the SNR.

8.8. EMERGING TECHNOLOGIES AND TRIGONOMETRY

Within the dynamic realm of technology, characterized by constant evolution and the primacy of invention, trigonometry assumes a steadfast role. In light of the imminent breakthroughs in several domains, including machine learning, quantum computing, and virtual reality (VR), trigonometry assumes a fundamental role as an enduring cornerstone that supports the fundamental structure of these revolutionary pursuits. Trigonometry's impact transcends its conventional domains, encompassing a wide range of phenomena such as quantum entanglement and immersive virtual worlds, therefore heralding a transformative era characterized by unprecedented possibilities (Lu et al., 2022).

8.8.1. Quantum Computing and Quantum Mechanics

The field of quantum computing, which involves the manipulation of the fundamental aspects of reality to perform computational tasks, may appear to be distant from the geometric origins of trigonometry. Upon conducting a more thorough analysis, the connection becomes evident. Quantum mechanics, the theoretical framework that serves as the foundation for quantum computing, incorporates a multitude of trigonometric concepts. The notion of wave-particle duality, which is a fundamental principle in the field of quantum physics, has similarities to trigonometric associations between frequency and waveforms.

Quantum entanglement is a phenomenon in which particles exhibit correlation such that the state of a single particle instantaneously affects the state of a different one, irrespective of the spatial separation between them. Trigonometric functions serve as a means to represent probability distributions and phases associated with entangled states, shedding light on the profound mathematical underpinnings of quantum events.

- **Example:** The Quantum State Vector:

In quantum mechanics, the quantum state of a system is represented by a state vector, often denoted as $|\psi$. Mathematically, this state vector can be expressed as a linear combination of basis states:

$$|\psi = \sum_i c_i |i$$

In this expression, $|\psi$ represents the quantum state of the system, $|i$ represents the basis states (eigenstates of a Hermitian operator), and c_i are complex coefficients that determine the probability amplitudes associated with each basis state. This mathematical expression is foundational in quantum mechanics and quantum computing. It encapsulates the idea that a quantum state can exist in a superposition of basis states, with the coefficients c_i representing the probability amplitudes of measuring the system in each of those basis states.

8.8.2. Artificial Intelligence and Data Analysis

Trigonometry assumes a crucial role in the domain of artificial intelligence (Bhatti et al., 2016), wherein robots endeavor to replicate human intellect. It serves as a fundamental tool for data processing and the identification of patterns. Machine learning algorithms, which are tasked with processing large volumes of data and generating predictions, frequently depend on trigonometric computations to unveil concealed patterns and correlations.

Consider a hypothetical scenario whereby an artificial intelligent (Bhatti et al., 2016) system is employed to analyze medical data with the objective of predicting illness outcomes. By utilizing trigonometry, the artificial intelligence system is able to discern nuanced relationships among different variables and construct models that inform medical judgments, leveraging mathematical principles to enhance the quality of patient treatment.

8.8.3. Virtual Reality and Immersive Experiences

The advent of VR technologies has brought about a significant transformation in our methods of engaging with digital settings, effectively submerging us in aesthetically appealing realms. Trigonometry plays a significant part in the development of VR, encompassing tasks such as producing three-dimensional (3D) images and tracking user motions. The mathematical principles governing angles, distances, and spins play a crucial role in ensuring the smooth alignment between virtual worlds and users' perceptual experiences.

The utilization of a VR headset enables individuals to be immersed in an immersive experience that transports them to a bygone era of human civilization. The trigonometric principles that regulate the tracking sensors of the headset provide precise translation of users' motions into the virtual environment, hence boosting the level of engagement and realism experienced by the users.

- **Example:** The Projection Formula in VR

In VR applications, especially when dealing with immersive 3D environments, trigonometric principles play a crucial role in calculating the projection of objects onto a 2D screen or viewport. One fundamental expression related to this is the projection formula, which involves trigonometric functions:

$$\left[x' = x \cdot \frac{d}{D}\right]$$

In this equation, x' represents the position of an object on the 2D screen, x is the object's position in 3D space, d is the distance from the viewer's eye to the screen (known as the viewer distance), and D is the distance from the viewer's eye to the object in 3D space. This mathematical expression illustrates how trigonometry is used to calculate the position of objects on a 2D screen in VR applications, ensuring that objects appear correctly from the viewer's perspective in immersive 3D environments.

8.8.4. 3D Printing and Geometric Modeling

In the domain of manufacturing, the utilization of 3D printing is revolutionizing the process of item fabrication. Trigonometry plays a significant role in modeling and designing procedures, providing guidance for the development of complex forms and structures. Geometric modeling, which relies on trigonometric calculations, guarantees the accurate replication of designs by 3D printers.

Consider the scenario of an aircraft engineer tasked with the construction of a very intricate turbine blade. Trigonometry empowers engineers to compute angles, parameters, and orientations that ensure optimal performance, underscoring the pivotal role of mathematics in shaping the future of manufacturing.

8.8.5. Biotechnology and Medical Imaging

Trigonometry plays a crucial role in facilitating advancements in healthcare imaging and diagnostics within the realm of biotechnology. Specifically, it provides essential assistance for the development and implementation of cutting-edge techniques like magnetic resonance imaging (MRI) and computed tomography or CT scans. These technologies utilize the fundamental concepts of trigonometry in order to rebuild intricate pictures of interior structures.

In the context of medical imaging, the role of a skilled technician who uses MRI to acquire intricate visual representations of a patient's cerebral structure is of key importance. Trigonometry-based methods are utilized to examine the interplay between the magnetic field and tissue characteristics, facilitating accurate imaging and diagnostic procedures.

- **Example:** The Trigonometric Resonance Imaging Formula

In MRI, which is widely used in medical imaging and biotechnology, trigonometric principles are essential for understanding the behavior of nuclear spins within a magnetic field.

One key concept is the Larmor frequency (ω_0), which represents the resonant frequency at which nuclear spins process in a magnetic field:

$$\omega_0 = \gamma B_0$$

In this equation, ω_0 represents the Larmor frequency, (γ) is the gyromagnetic ratio specific to the nucleus being imaged, and B_0 is the strength of the magnetic field.

8.8.6. Space Exploration and Celestial Navigation

Trigonometry maintains its relevance even in the most remote regions of outer space. Trigonometric concepts serve as the foundation for spacecraft navigating, interplanetary trajectory computations, and celestial navigation. Space explorers navigate through the universe by utilizing measurements of distances and angles between celestial entities.

Consider a hypothetical scenario in which a spaceship starts a mission to Mars. Trigonometry plays a crucial role in assisting mission planners with the computation of ideal trajectories, taking into consideration the effects of gravity and the orbits of celestial bodies, and assuring the achievement of successful interplanetary flight (Putro et al., 2023).

8.9. CHALLENGES AND FUTURE DIRECTIONS

Trigonometry, as an essential discipline within the field of mathematics, has exerted a substantial influence on the development of several technical disciplines. The applications of this technology range from navigation to electrical signal processing, encompassing fields such as architecture and VR, and have a widespread and significant influence. Nevertheless, with the continuous advancement of technology, the integration and growth of trigonometry in numerous disciplines present new problems and future perspectives.

8.9.1. Challenges

- **Complex Systems:** The study of complex systems involves the examination and analysis of intricate and interconnected phenomena. As technological advancements continue to progress, the field of trigonometry has difficulties in its ability to effectively represent and analyze sophisticated systems. The interaction among many variables and components can give rise to mathematical intricacies that necessitate the utilization of sophisticated computing techniques and algorithms.

- **Integration with Emerging Tech**: With the increasing significance of future technologies like quantum computers and artificial intelligence, the integration of trigonometry with these new paradigms becomes imperative. The application of mathematical principles to quantum computations or artificial intelligence models entails both prospects and obstacles.
- **Precision and Accuracy:** In disciplines such as navigation and astronomy, accuracy has utmost significance. In such instances, high degrees of precision are required while taking into consideration factors such as inaccuracies in measurements, variations in environmental circumstances, and constraints imposed by the instruments utilized.
- **Real-Time Applications:** Real-time decision-making is of utmost importance in businesses such as robots and driverless cars. Trigonometry encounters the task of delivering precise and efficient computations to steer these systems while minimizing the occurrence of delays or inaccuracies.

8.9.2. Future Directions

- **Quantum Trigonometry:** The investigation of the potential adaptation of trigonometry to quantum computation is a promising field, particularly in light of the increasing prominence of quantum computing. The utilization of quantum trigonometry has the potential to facilitate the development of more effective simulations for quantum systems and improve computational processes that include intricate wave functions.
- **Trigonometry in Augmented Reality:** The use of augmented reality (Johnson & Walker, 2011) software superimposes digital material onto the physical environment. The use of trigonometric principles has the potential to improve the mapping of space and the placement of virtual components with real-world environments, hence resulting in heightened levels of immersion.
- **Trigonometry in Biotechnology:** With the continuous progress of medical imaging and biotechnology, the utilization of

trigonometry has become indispensable in the analysis of intricate data obtained from modalities such as MRI and CT scans. The utilization of sophisticated algorithms grounded in trigonometric principles has the potential to enhance the precision of diagnostic procedures and therapeutic interventions.

- **Innovations in Signal Processing:** The uses of trigonometry in signal processing are expected to expand further due to the proliferation of IoT devices, the development of 5G networks, and other advancements in the field. In the coming years, there are prospects for the development of enhanced algorithms that can effectively handle extensive data sets with reduced energy usage.
- **Trigonometry in Sustainable Design:** Trigonometry plays a significant role in the realm of sustainable design within the fields of architecture and engineering. By using trigonometric principles, professionals may effectively enhance the energy efficiency, solar orientation, and material use of structures. This aligns with the increasing focus on environmentally responsible approaches to design and construction.
- **Education and Outreach:** The future trajectory of education and outreach endeavors should prioritize enhancing the accessibility and engagement of trigonometry for students and learners. The utilization of interactive software, simulations, and gamification has the potential to elucidate trigonometric ideas and cultivate a heightened interest in STEM disciplines.
- **Interdisciplinary Collaborations:** The future trajectory of trigonometry is contingent upon fostering interdisciplinary relationships. The multidisciplinary nature of technology facilitates collaboration among mathematicians, engineers, scientists, and artists, enabling the development of novel solutions that harness the versatile applications of trigonometry.

In summary, the obstacles, and prospective pathways of trigonometry in technology exemplify the dynamic terrain of advancement and investigation. The increasing complexity and variety of technology necessitate the ongoing relevance and fundamental importance of trigonometry in influencing our methods of navigation, communication, creation, and interaction with the surrounding environment. Through proactive engagement with obstacles and the proactive adoption of emerging prospects, the symbiotic relationship between mathematics and technology is positioned to facilitate revolutionary progress in several fields.

Remember

Trigonometry played a pivotal part during the Age of Discovery as it enabled sailors to traverse the vast expanses of the oceans by utilizing celestial objects such as stars and the sun.

8.10. SUMMARY

This chapter explores the significant influence of trigonometry across several technical fields. Trigonometry, an area of study within mathematics that is founded upon geometric connections and trigonometric functions, serves as a fundamental basis for a multitude of technological advancements that significantly influence our contemporary society. Trigonometry has a wide range of applications spanning several fields, such as navigation, architecture, sound, the internet, electronics, and other related domains.

Trigonometry plays a crucial role in the field of navigation or GPS technology by enabling precise location and navigating through the utilization of triangulation and the trilateration method concepts. The utilization of satellite signals allows gadgets to accurately ascertain their positions, hence altering the manner in which we traverse our surroundings.

Trigonometry serves as a fundamental basis in the field of music and sound technology, providing the necessary framework for the generation of waveforms, the harmonics, and harmonics that shape auditory perception. Trigonometry plays a crucial role in deepening our comprehension and manipulation of sound, encompassing activities such as the analysis of musical compositions and the creation of virtual soundscapes.

Trigonometry plays a fundamental role in the domains of circuits and signal processing, providing the necessary foundation for tasks such as data analysis, modification, and processing of digital signals. The effect of this phenomenon spans from the filtration of undesirable noise to the manipulation of signals for optimal transmission, hence enhancing human skills in communication and information processing.

Trigonometry plays a crucial role in the fields of architecture, engineering, and construction, providing essential guidance for tasks such as structural analysis, geometric planning, and aesthetic considerations. The utilization of geometry in architecture serves to maintain the structural integrity of buildings, facilitates precise quantification, and contributes to the aesthetic appeal of architectural masterpieces.

Trigonometry's influence is particularly evident in the domains of graphic design, animation, and augmented and mixed reality, where it imbues digital environments and animations with vitality. Trigonometry employs rotations, changes, and geometric modeling to engender visual experiences that engage our senses of sight and sound.

Trigonometry plays a crucial role in exploring new boundaries within developing technologies, including quantum information processing, artificial intelligence, and biotechnology. The adaptability of this technology is demonstrated in cutting-edge sectors, as it is capable of adapting to quantum physics, facilitating data processing, and providing guidance for medical imaging procedures.

The chapter also emphasizes the obstacles and prospective avenues for future exploration. With the increasing complexity and multidisciplinary nature of technology, it is imperative for trigonometry to evolve in order to effectively represent sophisticated systems and seamlessly interact with emerging technologies. Persistent challenges are observed in the areas of accuracy, applications in real-time, and seamless integration. Nevertheless, the future of trigonometry is promising since it encompasses other emerging fields such as quantum trigonometry, augmented reality applications, environmentally friendly design, and other potential advancements.

In conclusion, the chapter highlights the significant influence of trigonometry on several technical fields. Trigonometry serves as a potent and influential entity, amalgamating fundamental mathematical principles with inventive practical implementations, so molding our engagements within the realm of digital technology and moving mankind towards a future brimming with abundant prospects.

REFERENCES

1. Andreescu, T., & Feng, Z., (2005). Trigonometric fundamentals. *103 Trigonometry Problems: From the Training of the USA IMO Team,* 1–62.
2. Bandeira, J., & De Souza, R. M. C., (2006). *New Trigonometric Transforms Over Prime Finite Fields for Image Filtering* (Vol. 13, pp. 65–90). Paper presented at the 2006 International Telecommunications Symposium.
3. Bayeh, C. Z., & Mastorakis, N. E., (2013). Application of the rectangular trigonometry in industrial electronic systems with analyzing, modeling, and simulating the function rectangular rit. *Statistics, 12,* 13.
4. Bedogni, L., Gramaglia, M., Vesco, A., Fiore, M., Härri, J., & Ferrero, F., (2015). The Bologna ringway dataset: Improving road network conversion in SUMO and validating urban mobility via navigation services. *IEEE Transactions on Vehicular Technology, 64*(12), 5464–5476.
5. Bhatti, Z., Ismaili, I. A., Zardari, S., Malik, H. A. M., & Karbasi, M., (2016). Procedural animation of 3d humanoid characters using trigonometric expressions. *Bahria University Journal of Information & Communication Technologies (BUJICT), 9*(2), 1–40.
6. Blanquer, S. B., Werner, M., Hannula, M., Sharifi, S., Lajoinie, G. P., Eglin, D., & Grijpma, D. W., (2017). Surface curvature in triply-periodic minimal surface architectures as a distinct design parameter in preparing advanced tissue engineering scaffolds. *Biofabrication, 9*(2), 025001.
7. Brodetsky, S., (1919). *(1) Analytic Geometry (2) Elementary Mensuration, Constructive Plane Geometry, And Numerical Trigonometry (3) Lectures on the Philosophy of Mathematics* (Vol. 4, pp. 45–60). In: Nature Publishing Group UK London.
8. Brodetsky, S., (1924). *(1) Practical Mathematical Analysis (2) Arithmetic for Engineers: Including Simple Algebra, Mensuration, Logarithms, Graphs, Trigonometry, and the Slide Rule; with an Appendix on Verniers and Micrometers (3) Mathematics for Engineers (4) Line Charts for Engineers (5) Fundamentals of Practical Mathematics* (Vol. 13, pp. 655–800). In: Nature Publishing Group UK London.
9. Cetin, O. F., (2015). Students perceptions and development of conceptual understanding regarding trigonometry and trigonometric

function. *Educational Research and Reviews, 10*(3), 338–350.

10. Confrey, J., Hoyles, C., Jones, D., Kahn, K., Maloney, A. P., Nguyen, K. H., & Pratt, D., (2010). Designing software for mathematical engagement through modeling. *Mathematics Education and Technology-Rethinking the Terrain: The 17th ICMI Study* (pp. 19–45).
11. Cruz, F. R., Cruz, A. D., Fabricante, J., & Diza, H. M., (2020). Comparison of guitar and piano instruments in kiss the rain and bongga ka day songs using Fourier analysis. *Southeast Asian Journal of Science and Technology, 5*(1).
12. Cruz-Barroso, R., Daruis, L., González-Vera, P., & Njåstad, O., (2007). Sequences of orthogonal Laurent polynomials, bi-orthogonality, and quadrature formulas on the unit circle. *Journal of Computational and Applied Mathematics, 206*(2), 950–966.
13. Cundy, H. M., (1971). Fundamentals of mathematics from and advanced viewpoint. Vols. 3 and 4; geometry and geometric analysis; solid geometry and spherical trigonometry. By EG Kogbetliantz. Pp. xvii, 578.£ 14.75. 1969.(Gordon and Breach). (Volumes available separately in paperback,£ 4.25 each). *The Mathematical Gazette, 55*(394), 482–483.
14. Curri, E., (2012). *Using Computer Technology in Teaching and Learning Mathematics in an Albanian Upper Secondary School: The Implementation of SimReal in Trigonometry Lessons* (Vol. 17, pp. 1–90). Universitetet i Agder; University of Agder.
15. Ellaithy, D. M., El-Moursy, M. A., Ibrahim, G. H., Zaki, A., & Zekry, A., (2017). Double logarithmic arithmetic technique for low-power 3-D graphics applications. *IEEE Transactions on Very Large Scale Integration (VLSI) Systems, 25*(7), 2144–2152.
16. Giordano, M., & Wanderley, M. M., (2013). *Perceptual and Technological Issues in the Design of Vibrotactile-Augmented Interfaces for Music Technology and Media.* Paper presented at the Haptic and Audio Interaction Design: 8th International Workshop, HAID 2013, Daejeon, Korea, April 18-19, 2013, Revised Selected Papers 8.
17. Hosch, W., (2010). *The Britannica Guide to Algebra and Trigonometry*. Britannica Educational Publishing.
18. Huang, G. S., Hsu, P. C., & Kao, M. C., (2020). *Navigation Technology Using Inertial Elements to Compensate for GPS Signal-Shaded in Real Time* (Vol. 19, pp. 766–800). Paper presented at the 2020 5th

International Conference on Green Technology and Sustainable Development (GTSD).

19. Hwang, C., & Hwang, L. S., (2002). Use of geoid for assessing trigonometric height accuracy and detecting vertical land motion. *Journal of Surveying Engineering, 128*(1), 1–20.
20. Jackson, R., (2021). Visual illusions. In: *Encyclopedia of Evolutionary Psychological Science* (Vol. 3, No. 1, pp. 8433–8437). Springer.
21. Jelatu, S., Kurniawan, Y., Kurnila, V. S., Mandur, K., & Jundu, R., (2019). Collaboration TPS learning model and m-learning based on android for understanding of trigonometry concepts with different cognitive style. *International Journal of Instruction, 12*(4), 545–560.
22. Johnson, J. D., & Walker, M., (2011). Trigonometry students' knowing when to use handheld CAS technology to make sense of mathematics. *Journal of Mathematical Sciences and Mathematics Education, 6*(2), 17–33.
23. Jones, W. B., Njåstad, O., & Thron, W., (1989). Moment theory, orthogonal polynomials, quadrature, and continued fractions associated with the unit circle. *Bulletin of the London Mathematical Society, 21*(2), 113–152.
24. Kabaca, T., & Aktumen, M., (2010). Using GeoGebra as an expressive modeling tool: Discovering the anatomy of the cycloid's parametric equation. *GeoGebra The New Language for the Third Millennium, 1*(1), 63–82.
25. Kekre, H., & Solanki, J., (1978). Comparative performance of various trigonometric unitary transforms for transform image coding. *International Journal of Electronics Theoretical and Experimental, 44*(3), 305–315.
26. Kissane, B., & Kemp, M., (2009). *Teaching and Learning Trigonometry with Technology* (Vol. 23, No. 1, pp. 123–444). Paper presented at the 14th Asian Technology Conference in Mathematics.
27. Kouvonen, O., (2019). *Electronic Music Production Methods: Musical Electronic Hardware Instruments in Album Production Process, 15*(3), 127–340.
28. Landwehr, K., (2022). Geometrical visual illusions revisited: The curse of symmetry, the cure of sighting, and taxing task demands. *Symmetry, 14*(12), 2550.
29. Latibjonovich, E. S., (2022). Features of teaching trigonometry in

specialized schools. *Open Access Repository, 8*(9), 82–87.

30. Lee, S. K., Becker, A., Fekete, S. P., Kröller, A., & McLurkin, J., (2014). *Exploration Via Structured Triangulation by a Multi-Robot System with Bearing-Only Low-Resolution Sensors.* Paper presented at the 2014 IEEE International Conference on Robotics and Automation (ICRA), *12,* 13–90.
31. Lenahan, S., (2019). *The Precalculus Trigonometry Coordinator Plans to Incorporate Four to Six Lab Assignments Into Each Section of MA 113 Precalculus Trigonometry, 3*(1), 113–200.
32. Li, W., Li, W., Cui, X., Zhao, S., & Lu, M., (2018). A tightly coupled RTK/INS algorithm with ambiguity resolution in the position domain for ground vehicles in harsh urban environments. *Sensors, 18*(7), 2160.
33. Liang, P. C., & Krause, P., (2014). *Real-Time Indoor Patient Movement Pattern Telemonitoring with One-Meter Precision* (Vol. 7, pp. 322–400). Paper presented at the 2014 4th International Conference on Wireless Mobile Communication and Healthcare-Transforming Healthcare Through Innovations in Mobile and Wireless Technologies.
34. Liu, J., Lughofer, E., & Zeng, X., (2015). Aesthetic perception of visual textures: A holistic exploration using texture analysis, psychological experiment, and perception modeling. *Frontiers in Computational Neuroscience, 9*, 134.
35. Lostanlen, V., Andén, J., & Lagrange, M., (2019). Fourier at the heart of computer music: From harmonic sounds to texture. *Comptes Rendus. Physique, 20*(5), 461–473.
36. Loudon, J. C., (1845). *Self-instruction for Young Gardeners, Foresters, Bailiffs, Land-stewards, and Farmers: In Arithmetic and Book-keeping, Geometry, Mensuration, and Practical Trigonometry, Mechanics, Hydrostatics, and Hydraulics, Land-surveying, Levelling, Planning, and Mapping, Architectural Drawing, and Isometrical Projection and Perspective: With Examples, Showing Their Application to Horticultural and Agricultural Purposes* (Vol. 1, pp. 20–30).
37. Lu, W., Wu, Q., Zhang, J., Rao, J., Li, C., & Zheng, S., (2022). TANKbind: Trigonometry-aware neural networks for drug-protein binding structure prediction. *Advances in Neural Information Processing Systems, 35*, 7236–7249.
38. Melchels, F. P., Bertoldi, K., Gabbrielli, R., Velders, A. H., Feijen, J., & Grijpma, D. W., (2010). Mathematically defined tissue engineering

scaffold architectures prepared by stereolithography. *Biomaterials, 31*(27), 6909–6916.

39. Merched, R., (2006). On OFDM and single-carrier frequency-domain systems based on trigonometric transforms. *IEEE Signal Processing Letters, 13*(8), 473–476.
40. Moreolo, M. S., Sacchieri, V., & Cincotti, G., (2009). *Signal Processing Based on Trigonometric Transforms for High-Speed Optical Networks* (Vol. 12, No. 5, pp. 90–110). Paper presented at the 2009 11th International Conference on Transparent Optical Networks.
41. Mosese, N. M., (2017). *Evaluating the Effectiveness of the Use of Information and Communication Technology in the Teaching and Learning of Trigonometry Functions in Grade, 12*(17), 80–95.
42. Mosese, N., & Ogbonnaya, U. I., (2021). GeoGebra and students' learning achievement in trigonometric functions graphs representations and interpretations. *Cypriot Journal of Educational Sciences, 16*(2), 827–846.
43. Nam, B. G., Kim, H., & Yoo, H. J., (2007). A low-power unified arithmetic unit for programmable handheld 3-D graphics systems. *IEEE Journal of Solid-State Circuits, 42*(8), 1767–1778.
44. Ni, L., Jung, H. Y., & Zhou, Z., (2015). *Enhancing First-year Engineering Students' Trigonometry Learning Experience* (Vol. 2, No. 1, pp. 123–200). Paper presented at the 2015 ASEE Annual Conference & Exposition.
45. Osudin, D., (2022). *Real-Time Physics and Graphics Engine for non-Euclidean Geometry Using Spherical and Hyperbolic Trigonometry* (Vol. 8, pp. 30–55). City, University of London.
46. Park, T. H., (2009). *Introduction to Digital Signal Processing: Computer Musically Speaking* (Vol. 12, pp. 22–55). World Scientific.
47. Petrescu, R. V., Aversa, R., Apicella, A., Kozaitis, S., Abu-Lebdeh, T., & Petrescu, F. I., (2017). The inverse kinematics of the plane system 2-3 in a mechatronic MP2R system, by a trigonometric method. *Journal of Mechatronics and Robotics, 1*(2), 75–87.
48. Prabowo, A., Anggoro, R., Adiyanto, R., & Rahmawati, U., (2018). *Interactive Multimedia-Based Teaching Material for Trigonometry.* Paper presented at the Journal of Physics: Conference Series, *2*(1), 112–200.
49. Putro, B. S., Selasdini, V., & Saputra, I. R., (2023). *Analysis the*

Importance of Applying Celestial Navigation to Avoid the Deficiency During Ship Inspections at LPG/C Navigator Aries (Vol. 5, No. 1, pp. 54–70). Paper presented at the AIP Conference Proceedings.

50. Qiao, H., Wen, S., Wu, L., & Zeng, B., (2022). Research on wind power project investment risk evaluation based on fuzzy-gray clustering trigonometric function. *Energy Reports, 8*, 1191–1199.
51. Reid, M., Menten, K., Zheng, X., Brunthaler, A., Moscadelli, L., Xu, Y., & Hirota, T., (2009). Trigonometric parallaxes of massive star-forming regions. VI. Galactic structure, fundamental parameters, and noncircular motions. *The Astrophysical Journal, 700*(1), 137.
52. Rolls, E. T., (2021). Neurons including hippocampal spatial view cells, and navigation in primates including humans. *Hippocampus, 31*(6), 593–611.
53. Serpe, A., & Frassia, M. G., (2021). Artefacts teach-math. The meaning construction of trigonometric functions. *Atti Della Accademia Peloritana dei Pericolanti-Classe di Scienze Fisiche, Matematiche e Naturali, 99*(S1), 15.
54. Shaghaghian, Z., Burte, H., Song, D., & Yan, W., (2022). *Learning Spatial Transformations and Their Math Representations Through Embodied Learning in Augmented Reality* (Vol. 7, No. 1, pp. 55–98). Paper presented at the International Conference on Human-Computer Interaction.
55. Sparks, J. L., (2023). *A Visual Illusion In an+ bn= cn and ax+ by= cz? The Fermat's Last Theorem and the Beal Conjecture Specialty Equations and the Birch-Swinnerton-Dyer Conjecture* (Vol. 13, No. 1, pp. 47–100). Walden University.
56. Strauß, R., (2005). *President of the "Arbeitskreis Triangulation of the "Arbeitsgemeinschaft der Vermessungsverwaltungen der Länder der Bundesrepublik Deutschland."* Paper presented at the GPS-Techniques Applied to Geodesy and Surveying: Proceedings of the International GPS-Workshop Darmstadt.
57. Suherman, S., Komarudin, K., Rosyid, A., Aryanita, S., Asriyanto, D., Aradika, P. T., & Anggoro, T., (2018). *Improving Trigonometry Concept Through STEM (Science, Technology, Engineering, and Mathematics) Learning* (Vol. 13, pp. 340–560). Paper presented at the International Conference on Multidisciplinary Academic (ICMA).
58. Trinh, C., Zlatanov, D., Zoppi, M., & Molfino, R., (2015). *A Geometrical*

Approach to the Inverse Kinematics of 6r Serial Robots with Offset Wrists (Vol. 2, No. 1, pp. 23–50). Paper presented at the International Design Engineering Technical Conferences and Computers and Information in Engineering Conference.

59. Vaitheeswaran, S., Venkatesh, M. S. K., Namitha, B., & Ravindra, V. C., (2012). *Aircraft Attitude Estimation from a Video Sequence Using Simple Trigonometric Expressions: GPS Outage, 1*(3), 56–79.
60. Van, B. G., (2009). *The Mathematics of the Heavens and the Earth: The Early History of Trigonometry* (Vol. 1, No. 1, pp. 1–20). Princeton University Press.
61. Van, I. B., (1914). Use of text books in the examination in trigonometry. *The Mathematics Teacher, 7*(1), 11–14.
62. Vaseghi, S. V., (2007). *Multimedia Signal Processing: Theory and Applications in Speech, Music, and Communications* (Vol. 52, pp. 133–222). John Wiley & Sons.
63. Weber, K., (2008). Connecting research to teaching: Teaching trigonometric functions: Lessons learned from research. *The Mathematics Teacher, 102*(2), 144–150.
64. Weng, T. S., Hsu, M. H., & Yang, D. C., (2017). A study investigating the use of 3D computer animations of trigonometric functions to enhance spatial perception ability. *International Journal of Information and Education Technology, 7*(1), 23.
65. Wildberger, N., (2015). *An Introduction to Rational Trigonometry and Chromogeometry* (Vol. 12, No. 23, pp. 45–61).
66. Wilhelm, J. A., (2002). *Assessing Student Understanding of Sound Waves and Trigonometric Reasoning in a Technology-Rich, Project-Enhanced Environment*. The University of Texas at Austin.
67. Wilson, S. J., (2008). Dynamic web tools for trigonometry. *International Journal of Continuing Engineering Education and Life Long Learning, 18*(5, 6), 698–707.
68. Yang, B., Dong, H., & El Saddik, A., (2017). Development of a self-calibrated motion capture system by nonlinear trilateration of multiple Kinects v2. *IEEE Sensors Journal, 17*(8), 2481–2491.
69. Yeshurun, S., (1982). The angle: A logical gap in teaching geometry and trigonometry and its remedy. *International Journal of Mathematical Education in Science and Technology, 13*(2), 133–138.
70. Zengin, Y., Furkan, H., & Kutluca, T., (2012). The effect of dynamic

mathematics software GeoGebra on student achievement in teaching of trigonometry. *Procedia-Social and Behavioral Sciences, 31*, 183–187.

71. Zhao, D., Huang, Z., Li, H., Chen, J., & Wang, P., (2017). An improved EEMD method based on the adjustable cubic trigonometric cardinal spline interpolation. *Digital Signal Processing, 64*, 41–48.
72. Ziatdinov, R., Yoshida, N., & Kim, T. W., (2012). Analytic parametric equations of log-aesthetic curves in terms of incomplete gamma functions. *Computer Aided Geometric Design, 29*(2), 129–140.

INDEX

D

E

O

P

Q

R

S

V

W